New Advances on Nutrients Recovery from Agro-Industrial and Livestock Wastes for Sustainable Farming

Editors

Mirko Cucina
Luca Regni

MDPI • Basel • Beijing • Wuhan • Barcelona • Belgrade • Manchester • Tokyo • Cluj • Tianjin

Editors
Mirko Cucina
Agricultural and Environmental
Sciences Department
University of Milan
Milan
Italy

Luca Regni
Agricultural, Food and
Environmental Sciences
Department
University of Perugia
Perugia
Italy

Editorial Office
MDPI
St. Alban-Anlage 66
4052 Basel, Switzerland

This is a reprint of articles from the Special Issue published online in the open access journal *Agronomy* (ISSN 2073-4395) (available at: www.mdpi.com/journal/agronomy/special_issues/ nutrient_recovery_sustainable_farming).

For citation purposes, cite each article independently as indicated on the article page online and as indicated below:

LastName, A.A.; LastName, B.B.; LastName, C.C. Article Title. *Journal Name* **Year**, *Volume Number*, Page Range.

ISBN 978-3-0365-2565-5 (Hbk)
ISBN 978-3-0365-2564-8 (PDF)

Contents

Editorial

New Advances on Nutrients Recovery from Agro-Industrial and Livestock Wastes for Sustainable Farming

Mirko Cucina [1,*] and Luca Regni [2]

[1] Gruppo Ricicla Lab., DiSAA, Università degli Studi di Milano, Via Celoria, 2-20133 Milano, Italy
[2] Research Unit of Arboriculture, Department of Agricultural, Food and Environmental Sciences, University of Perugia, Borgo XX Giugno, 74-06121 Perugia, Italy; luca.regni@unipg.it
*   Correspondence: mirko.cucina@unimi.it; Tel.: +39-3338579190

Citation: Cucina, M.; Regni, L. New Advances on Nutrients Recovery from Agro-Industrial and Livestock Wastes for Sustainable Farming. *Agronomy* **2021**, *11*, 2308. https://doi.org/10.3390/agronomy11112308

Received: 26 October 2021
Accepted: 8 November 2021
Published: 15 November 2021

Publisher's Note: MDPI stays neutral with regard to jurisdictional claims in published maps and institutional affiliations.

The world's population continues to rise, with a medium-variant forecast predicting that by 2050, the global population will have surpassed 10 billion people [1]. As a result, it is clear that there is a need for increased food production order to meet the world's expanding population as well as to meet the Sustainable Development Goals, i.e., zero Hunger, which was defined by the United Nations in 2015 [2]. Since agriculture is the primary source of food [3], improving crop yields is a top concern. Agriculture productivity has mostly been improved via the use of fertilizers in recent decades, and global demand for nitrogen, phosphate, and potassium for fertilizer usage is predicted to grow by nearly 10% from 2016 to 2022 [4]. The growing need for fertilizers raises major concerns, which are mostly about their production and the environmental impacts related to their production and use. Among the most used synthetic fertilizers, N and P are frequently obtained from nonrenewable resources that use high-cost methods [5,6], and the environmental concerns associated with their usage are well documented (e.g., eutrophication, gaseous emissions) [7]. Biobased fertilizers (BBFs) have improved the sustainability of agriculture by reducing the use of non-renewable resources and the impact of agriculture on the environment [8]. Agro-industrial and animal husbandry waste might provide a sustainable supply of BBFs, which would be low-cost and ecologically beneficial. Different technologies for recovering nutrients from organic waste are available; however, they are not widely used. The primary obstacles to nutrient recovery from agro-industrial and animal waste include unfavorable waste stream features (e.g., the presence of organic contaminants), technical obstacles (i.e., obtaining sanitized products), and a lack of information about the quality and efficacy of BBFs.

In this context, the aim of this Special Issue of *Agronomy*, "New Advances on Nutrients Recovery from Agro-Industrial and Livestock Wastes for Sustainable Farming", was to advance knowledge on (i) the analysis of agro-industrial and livestock waste streams and potential for nutrient recovery and supply, (ii) technologies for nutrient recovery, (iii) quality of biobased fertilizers, (iv) laboratory and field assessment of biobased fertilizers, and (v) future challenges in nutrient recovery.

Five research articles have been published in this Special Issue. These research papers cover a wide variety of research areas that are related to nutrient recovery from animal and agro-industrial wastes, including process optimization and improvement, the effects of BBFs on soil fertility and plant growth, and comparison of different organic farming processes.

Process optimization is crucial to ensuring the production of high-value BBFs, and this aspect was investigated in one of the papers published in this Special Issue [9]. In this study, it was highlighted that the evolution of the compost process and compost quality was strongly related to the content of a soluble form of nitrogen (i.e., water-extractable N, WEN) in the composting mixtures. This observation was used to propose new parameters (i.e., WEN, TOC/WEN) to optimize the composting process.

An improved vermicomposting system was proposed to treat dairy wastewater, rice, straw, and cow manure and to recover nutrients [10]. The results of this study showed that this can represent a feasible method for the simultaneous disposal of organic wastes, especially in poor countries, because it incurs lower costs and has a lower impact on the environment.

The effects of compost and compost tea (CT) applications on soil fertility, plant growth, and the environment were evaluated in two research articles [11,12]. Dairy manure compost and food waste compost showed different effects on soil microbial community composition, but in both cases, the results suggested that recycled waste composts contribute to biologically based nitrogen cycling and can increase tree growth, especially within the first year after application [11]. CT application induced plant growth and defense in pepper plants against *Phytophthora capsici* and *Rhizoctonia solani* because of its relevant soluble nutrient content and microbiota richness, which provided a novel point for plant nutrition and protection in horticultural crops [12].

Finally, the results of a comparison between different BBF applications on soil in organic farming showed that the addition of organic matter improved crop yields regardless of its source [13].

This Special Issue of *Agronomy*, "New Advances on Nutrients Recovery from Agro-Industrial and Livestock Wastes for Sustainable Farming", contains different papers presenting new results concerning different aspects of nutrient recovery and utilization from agro-industrial and animal waste. These research papers cover a variety of biological processes that are related to nutrient recovery (i.e., composting, vermicomposting), including the evaluation of the effects of BBF applications on soil systems, plant growth, and the environment.

The Academic Editors of this Special Issue (Dr. Mirko Cucina and Dr. Luca Regni) hope that this collection of research articles will stimulate research in the field of nutrient recovery from organic waste, not only to improve knowledge but also for applications in sustainable farming.

Author Contributions: Conceptualization, M.C. and L.R.; writing—original draft preparation, M.C.; writing—review and editing, L.R. All authors have read and agreed to the published version of the manuscript.

Funding: This research received no external funding.

Conflicts of Interest: The authors declare no conflict of interest.

Abbreviations

BBFs	Biobased fertilizers
CT	Compost-tea
N	Nitrogen
TOC	Total Organic Carbon
WEN	Water-extractable Nitrogen

References

1. ONU. World Population Prospects 2019: Highlights. Department of Economic and Social Affairs, Population Division. 2019. Available online: https://population.un.org/wpp/Publications/Files/WPP2019_Highlights.pdf (accessed on 11 November 2021).
2. The United Nations. Take Action for the Sustainable Development Goals—United Nations Sustainable Development. United Nations Sustainable Development. 2020. Available online: https://www.un.org/sustainabledevelopment/sustainable-development-goals/ (accessed on 18 October 2021).
3. FAO. Food Outlook—Biannual Report on Global Food Markets. 2020. Available online: https://doi.org/10.4060/cb1993en (accessed on 18 October 2021). [CrossRef]
4. FAO. World Fertilizer Trends and Outlook to 2022. 2020. Available online: https://doi.org/10.4060/ca6746en (accessed on 18 October 2021). [CrossRef]
5. Cherkasov, N.; Ibhadon, A.O.; Fitzpatrick, P. A review of the existing and alternative methods for greener nitrogen fixation. *Chem. Eng. Process.* **2015**, *90*, 24–33. [CrossRef]
6. Günther, S.; Grunert, M.; Müller, S. Overview of recent advances in phosphorus recovery for fertilizer production. *Eng. Life Sci.* **2018**, *18*, 434–439. [CrossRef] [PubMed]
7. Khan, M.N.; Mohammad, F. Eutrophication: Challenges and solutions. In *Eutrophication: Causes, Consequences and Control*; Springer: Dordrecht, Germany, 2014; pp. 1–15. [CrossRef]

8. Wang, Y.; Zhu, Y.; Zhang, S.; Wang, Y. What could promote farmers to replace chemical fertilizers with organic fertilizers? *J. Clean. Prod.* **2018**, *199*, 882–890. [CrossRef]
9. Pezzolla, D.; Cucina, M.; Proietti, P.; Calisti, R.; Regni, L.; Gigliotti, G. The Use of New Parameters to Optimize the Composting Process of Different Organic Wastes. *Agronomy* **2021**, *11*, 2090. [CrossRef]
10. Liu, X.; Geng, B.; Zhu, C.; Li, L.; Francis, F. An Improved Vermicomposting System Provides More Efficient Wastewater Use of Dairy Farms Using *Eisenia fetida*. *Agronomy* **2021**, *11*, 833. [CrossRef]
11. Hodson, A.K.; Sayre, J.M.; Lyra, M.C.C.P.; Rodrigues, J.L.M. Influence of Recycled Waste Compost on Soil Food Webs, Nutrient Cycling and Tree Growth in a Young Almond Orchard. *Agronomy* **2021**, *11*, 1745. [CrossRef]
12. González-Hernández, A.I.; Suárez-Fernández, M.B.; Pérez-Sánchez, R.; Gómez-Sánchez, M.Á.; Morales-Corts, M.R. Compost Tea Induces Growth and Resistance against *Rhizoctonia solani* and *Phytophthora capsici* in Pepper. *Agronomy* **2021**, *11*, 781. [CrossRef]
13. Antošovský, J.; Prudil, M.; Gruber, M.; Ryant, P. Comparison of Two Different Management Practices under Organic Farming System. *Agronomy* **2021**, *11*, 1466. [CrossRef]

 *agronomy*

Article

The Use of New Parameters to Optimize the Composting Process of Different Organic Wastes

Daniela Pezzolla [1,†], Mirko Cucina [2,*,†], Primo Proietti [3], Roberto Calisti [3], Luca Regni [3] and Giovanni Gigliotti [1]

[1] Department of Civil and Environmental Engineering, University of Perugia, Via G. Duranti, 93-06125 Perugia, Italy; daniela.pezzolla@unipg.it (D.P.); giovanni.gigliotti@unipg.it (G.G.)
[2] Gruppo Ricicla Lab.—DiSAA—Università degli Studi di Milano, Via Celoria, 2-20133 Milano, Italy
[3] Department of Agricultural, Food and Environmental Sciences, University of Perugia, Research Unit of Arboriculture, Borgo XX Giugno, 74-06121 Perugia, Italy; primo.proietti@unipg.it (P.P.); saacr@libero.it (R.C.); luca.regni@unipg.it (L.R.)
[*] Correspondence: mirko.cucina@unimi.it; Tel.: +39-333-857-9190
[†] Authors contributed equally to this work.

Abstract: The correct development of the composting process is essential to obtain a product of high value from organic wastes. Nowadays, some composting mixture parameters (i.e., air-filled porosity, moisture and the C/N ratio) are used to optimize the composting process, but their suitability is still debated because the literature reports contrasting results. This paper aimed to find other parameters that control the correct development of composting. The relationship between these and the compost quality was then verified. Twelve different composting mixtures were prepared using different organic wastes and bulking agents and were aerobically treated in a 300 L composter. The physico-chemical and chemical parameters of initial mixtures were analyzed, with particular regard to the total and water-extractable forms of organic C and N and their ratios and correlated with the temperature measured during composting. A positive correlation between temperature parameters during the active phase and soluble forms of N in the initial mixtures was found. A high total organic C to soluble N ratio in the composting mixtures was correlated with the low quality of the compost produced. Based on the results, a minimum content of WEN (water-extractable N) (0.4% w/w) or a TOC/WEN (total organic C/WEN) ratio in the range of 40–80 was recommended to ensure the correct development of the process and to produce compost of high quality.

Keywords: biological treatments; biomass; nutrients recovery; recycle; waste treatment

Citation: Pezzolla, D.; Cucina, M.; Proietti, P.; Calisti, R.; Regni, L.; Gigliotti, G. The Use of New Parameters to Optimize the Composting Process of Different Organic Wastes. *Agronomy* **2021**, *11*, 2090. https://doi.org/10.3390/agronomy11102090

Academic Editor: Andrea Baglieri

Received: 23 September 2021
Accepted: 15 October 2021
Published: 19 October 2021

Publisher's Note: MDPI stays neutral with regard to jurisdictional claims in published maps and institutional affiliations.

1. Introduction

Municipal, agro-industrial and livestock activities generate large volumes of organic wastes, whose management can produce adverse effects on the environment (soil, water and atmosphere pollution). Composting represents a suitable and environmentally friendly disposal strategy for organic waste management. Composting aerobically degrades organic wastes to compost and two main by-products, heat and carbon dioxide [1,2]. Composting is a self-heating process that proceeds through three main steps: (1) The mesophilic phase (25–40 °C), (2) the thermophilic phase (55–65 °C) and (3) the curing phase. During composting, labile organic matter is mineralized and complex recalcitrant materials tend to concentrate, increasing the organic matter stabilization in the compost [3]. Compost is a nutrient-rich organic amendment able to provide N, P, K and organic matter to the soil, also contributing to soil C sequestration [2,4].

Composting has several advantages, such as (1) sanification from pathogens and weed seeds, (2) volume and odor reduction, (3) microbial stabilization, (4) disposal cost reduction and (5) the production of organic fertilizer (compost) with an economical added value, that can be used to replace chemical fertilizers when characterized by high-quality [5,6].

Conversely, gas emissions, large area requirements and net energy consumption were reported to be the main disadvantages of composting [6–8].

The correct development of composting throughout its active and curing phase is essential to obtain high-quality compost. During the active phase, the temperature of the biomass is one of the most significant parameters for evaluating process effectiveness, as it depends on the aerobic nature of the process. In addition, compost sanitation and biodegradation rates rely on temperature changes during composting, and Cáceres et al. introduced the degree-hour concept to composting to optimize the exposure of feedstock to high and intermediate temperatures throughout the entire process [9]. An effective composting process also allows for a high degree of stabilization and maturity [10]. Stability refers to the degree of organic matter decomposition, whereas maturity refers to the removal of phytotoxic compounds from the composting feedstock. Compost stability and maturity can be assessed using different parameters, as reviewed by Bernal et al. [5]. Stability can be evaluated by using respirometric methods or by studying organic matter transformations during composting. Maturity can be assessed using physical, chemical (C/N ratio, pH, electrical conductivity, humification indices) or biological (phytotoxicity) criteria [5].

Different organic materials (e.g., agricultural and agro-industry by-products, livestock wastes, sewage sludge, organic fraction of the municipal solid wastes) can be treated through composting and they are generally mixed with bulking materials to ensure the optimal conditions for microbial growth and process development [4]. Mixture preparation is essential for balancing moisture, pH and the C/N ratio for adequate aeration and microbial growth [11]. Although mixture formulation is commonly carried out to adjust the C/N ratio and moisture in their ideal range (20–25 and 40–60%, respectively), other physical parameters (e.g., porosity, filled air space) are important in composting mixtures [12–14]. In addition, composting may be performed at low or high C/N ratios (<20 and >30) as total C and total N may include recalcitrant fractions depending on the chemical characterization of organic material [1], e.g., fiber content [15]. Puyuelo et al. demonstrated that the biodegradable C/N ratio was significantly different from the C/N ratio in several organic wastes, and pointed out that the C/N ratio used to optimize biological treatments should be defined in biodegradable terms [16]. This is in accordance with Trémier et al. who affirmed that mixture formulation must consider the biodegradability of wastes rather than other parameters [14]. As a consequence, it is often reported in the literature that composting mixtures characterized by the optimal C/N ratio did not show correct process development, resulting in a low-temperature profile and/or low-quality composts [11,17,18].

Since biochemical transformations of organic matter during composting occur in the aqueous phase, mixture formulation based on water-extractable C and N, and their ratio, appear to be appropriate. As demonstrated by Said-Pullicino et al., the water-soluble organic matter is the most accessible fraction of organic matter for the microorganisms [19], and this can be useful for monitoring the behavior of the composting process. Nevertheless, there are no studies focused on the relationship between these two parameters (water-extractable C and water-extractable N) and the evolution of the process. In this context, the present study aimed to assess chemical parameters related to C and N for the formulation of composting mixtures. The selected parameters were related to both the development of composting and to the quality (stabilization and maturity) of the produced compost.

2. Materials and Methods

2.1. Organic Materials

Table 1 shows the organic materials used in the present study and their main chemical characteristics. The composting experiments were performed in order to reuse different organic putrescible materials as the organic fraction of the municipal solid waste, pig slurry, sewage sludge and olive mill wastes. Even digestates, derived from different anaerobic trials, were used as substrates for the composting process (anaerobically digested, AFB1-contaminated chopped corn and anaerobically digested pharmaceutical wastewater). All these organic putrescible materials were co-composted with bulking agents, in order to establish the recipe structure of starting mixtures, exposed particle surface area and porosity [14] for ensuring the optimal development of composting, and hence the quality of the final compost. In this study, the bulking agents were wood chips, (WC, particle size < 20 mm), broadleaf tree pruning (BLTP), conifer tree pruning (CTP) and chopped-tree pruning (particle size < 50 mm). The use of tree pruning at different particle sizes was considered for balancing the surface area for microbial biodegradation and, hence, to optimize the porosity [5]. Cereal straw (CS) was added as an absorbent agent, when appropriate.

Table 1. Main characteristics of raw materials and bulking agents used in the composting experiments.

Type	Acronym	Description	Moisture (%)	TOC (%)	TKN (%)	pH	Bulk Density (kg L^{-1})
Raw wastes	OFMSW	Organic fraction of the municipal solid wastes	69.7 ± 0.2	22.2 ± 0.1	1.3 ± 0.1	3.2 ± 0.0	0.8 ± 0.1
	PS	Pig slurry	93.4 ± 0.6	40.2 ± 0.3	3.7 ± 0.2	7.3 ± 0.1	1.0 ± 0.0
	CC	AFB1 contaminated chopped corn	14.0 ± 0.1	37.5 ± 0.0	1.5 ± 0.1	6.1 ± 0.1	0.4 ± 0.1
	AD-CC	Anaerobically digested AFB1 contaminated chopped corn	93.8 ± 0.1	21.8 ± 0.1	6.3 ± 0.3	7.5 ± 0.0	0.9 ± 0.1
	OMW	Olive mill wastes	65.1 ± 1.2	39.9 ± 0.5	0.9 ± 0.2	5.9 ± 0.0	0.8 ± 0.1
	SS	Sewage sludge	89.9 ± 0.3	18.7 ± 0.6	3.8 ± 0.1	8.0 ± 0.1	0.8 ± 0.1
	AD-PW	Anaerobically digested pharmaceutical wastewater	96.2 ± 0.2	41.4 ± 0.2	9.4 ± 0.4	7.4 ± 0.2	0.9 ± 0.0
Bulking agents	CS	Cereal straw	14.2 ± 0.0	43.2 ± 0.2	0.7 ± 0.0	6.7 ± 0.0	0.1 ± 0.0
	WC	Wood chips	57.2 ± 0.9	42.1 ± 0.2	0.6 ± 0.1	6.8 ± 0.1	0.4 ± 0.1
	BLTP	Broadleaf tree pruning	6.1 ± 0.0	34.4 ± 0.0	0.8 ± 0.1	5.7 ± 0.1	0.2 ± 0.0
	C-BLTP	Chopped broadleaf tree pruning	6.9 ± 0.0	33.9 ± 0.5	1.0 ± 0.2	6.1 ± 0.0	0.2 ± 0.0
	CTP	Conifer tree pruning	26.5 ± 0.5	39.6 ± 0.5	0.6 ± 0.1	6.8 ± 0.0	0.3 ± 0.1
	C-CTP	Chopped conifer tree pruning	24.8 ± 0.3	38.7 ± 0.7	0.7 ± 0.1	6.8 ± 0.1	0.3 ± 0.0

Mean value ± SEM, n = 3; TOC: Total organic C; TKN: Total Kjeldahl N, AFB1: Aflatoxin B1. Data are expressed on dry basis.

Raw materials and bulking and/or absorbent agents were collected in the area of Perugia (Umbria Region, Central Italy). Pig slurry, olive mill wastes and cereal straw were obtained from local farmers, whereas organic fractions of municipal solid wastes and sewage sludge were obtained from local companies that collect and treat wastes or wastewater, respectively. Digestates were obtained from pilot anaerobic digesters treating aflatoxin B1 (AFB1)-contaminated corn or pharmaceutical wastewaters. Bulking and absorbent materials were collected from a local plant nursery.

Representative samples of all organic materials were collected and portioned in two fractions. The first one was stored at 4 °C before the beginning of the experiments, while the second one was dried, crushed, sieved (<0.5 mm) and mixed for analytical determinations.

2.2. Composting Experiments

Twelve different composting mixtures were studied in this work to assess new parameters suitable for mixture formulation and composting optimization. The parameters chosen were used to explain both the progress of composting and the quality of the final compost. The five mixtures were those described in Cucina et al. [20], Cucina et al. [3] and Tacconi et al. [21], and different compositions were included to ensure that the studied parameters described in the following paragraphs could be suitable for a broad type of mixtures. Bulking and/or absorbent agents were mixed with raw materials to maximize the amount of organic waste treated and to adjust the physical properties (bulk density, air-filled porosity and moisture content), avoid water leaching and ensure optimal air circulation. The C/N ratio in the mixtures was not adjusted to the optimal values, since it was one of the parameters selected to be evaluated for its relationship with composting development and final compost quality. Based on these considerations, mixtures were prepared according to the proportions (fresh weight basis) reported in Table 2. Mixtures were placed in experimental plastic composters (0.6 m × 0.6 m × 0.8 m = 0.288 m^3) at outdoor conditions. All the experiments were performed through static composting, and passive aeration was maintained thanks to the porosity of the mixture itself. The temperature was measured once a day (same time of the day) in the center of the pile using temperature probes (Stainless Steel Temperature Probe, Vernier, Beaverton, OR, USA).

Table 2. Mixture compositions.

Mixture	Composition (% *w/w*)
1 [a]	43% OFMSW + 43% WC + 14% CC
2 [a]	40% PS + 40% WC + 12% CC + 8% CS
3	42% AD-CC + 54% WC + 4% CS
4	55% PS + 45% BLTP
5	55% PS + 45% CTP
6	55% PS + 45% C-BLTP
7	55% PS + 45% C-CTP
8	80% OMW + 20% C-BLTP
9	80% OMW + 20% C-CTP
10 [b]	70% SS + 30% WC
11 [b]	45% SS + 45% WC + 10% CS
12 [c]	50% AD-PW + 40% WC + 10% CS

[a] [21]; [b] [3]; [c] [20]. OFMSW: Organic fraction of the municipal solid wastes, CC: AFB1-contaminated chopped corn, PS: Pig slurry, AD-CC: Anaerobically digested AFB1-contaminated chopped corn, OMW: Olive mill wastes, SS: Sewage sludge, AD-PW: Anaerobically digested pharmaceutical wastewater, CS: Cereal straw, WC: Wood chips, BLTP: Broadleaf tree pruning, CTP: Conifer tree pruning, C-BLTP: Chopped broadleaf tree pruning, C-CTP: Chopped conifer tree pruning.

The intensity of the composting process in the active phase was evaluated by calculating the time that the mixture temperature was higher than 55 °C (h > 55 °C) and the cumulative degree hours (DH), as described by Cáceres et al. [9]. The active phase of composting was considered completed when a stable temperature was reached (about 30 days). At the end of the active phase, the mixtures were sieved to 20 mm, placed in open boxes without aeration and mixed once a week for 60 days (curing phase). Representative samples of the initial mixtures and mature compost were collected and portioned in two fractions: The first one was stored at 4 °C for the determination of water-extractable C and N, while the second one was dried, crushed, sieved (<0.5 mm) and mixed for analytical determinations.

2.3. Analytical Methods

Moisture was determined on fresh samples by drying at 105 °C to constant weight. Total organic carbon (TOC), total Kjeldahl-N (TKN) and ammonium–N were analyzed according to standard methods [22]. pH and electrical conductivity were analyzed after water extraction of fresh samples (solid to water ratio of 1:10 *w/w*) using a glass electrode (pH-Meter Basic 20+, Crison Instruments, Barcelona, Spain) and a conductivity cell (Ec-Meter Basic 30+, Crison Instruments, Barcelona, Spain), respectively. Bulk density was determined following standard procedures [23], and air-filled porosity (AFP) was calculated from the wet bulk density as suggested by Alburquerque et al. [24]. Organic matter (OM) loss was calculated as described by Gigliotti et al. [25]. Humic-like substances were extracted and purified as described by Ciavatta et al. [26] and C quantification in the extracts was carried out using high-temperature combustion (805 °C, Pt catalyzed) followed by CO_2 infrared detection (Analyzer multi N/C 2100S, Analytic Jena, Überlingen, Germany). A germination assay employing cress seeds was used for the determination of potential phytotoxicity in mature compost [20].

Water-extractable organic C (WEOC) and N (WEN) were extracted from fresh samples (solid to water ratio 1:10 *w/w*) on a horizontal shaker for 24 h at room temperature. The suspensions were then centrifuged at 8000 rpm for 12 min and filtered (0.45 μm). Extracts were analyzed for organic C and N content using Pt-catalyzed, high-temperature combustion (805 °C) followed by infrared detection of CO_2 and chemiluminescence detection of NO. Water-extractable organic N (WEON) was determined by taking the difference between WEN and ammonium-N.

2.4. Statistics

Results are the arithmetic mean of three replicates, and the standard error of the mean (SEM) was reported along with the average value (Microsoft Excel Software). Bivariate correlation tests (Pearson's linear correlation coefficient) were used to study the associations between composting development and compost quality and the selected parameters of the mixtures (Microsoft Excel Software) (significance of $p < 0.01$ and $p < 0.05$, n = 12). Data passed the requisite tests for running a linear regression (plot of the data reveals a linear relationship, residuals are statistically independent, residuals are homoscedastic and residuals are unbiased).

3. Results and Discussion

3.1. Organic Materials Characteristics and Temperature Parameters

In the present work, the main characteristics of all raw materials used in the study (Table 1) were related to the correct development of the active phase of composting.

The correct physical structure of composting mixtures permits the oxygen availability for microorganisms and is essential for mineralization phenomena, and the oxygen availability mostly depends on the bulk density and moisture of the starting mixture [14,27]. Concerning the bulk density, the lowest values were observed, as expected, for the ligno-cellulosic materials (WC, BLTP, CTP, CS), and the chopped tree pruning did not increase this parameter. Among the bulking agents, the WC and the CTP were characterized by the highest values of the C/N ratio (70.2 and 66.0, respectively) and were used in the trials in a range from 20 to 58% w/w. In a study carried out by Barrena et al. [12], it was observed that the raw pruning wastes are also characterized by a high C/N ratio (42 and 52), and the semi-composted pruning resulted in a similar C/N ratio. All the raw wastes studied in the present work were characterized by a lower C/N ratio than the lignocellulosic materials, and this was particularly true for both digestate and sewage sludge, which showed values < 5. It is important to know the C/N ratio because it represents an indicator of nutritional balance for microbial activity [5]. In this case, the use of a large amount of pig slurry, anaerobic digestate and sewage sludge can result in a C/N ratio lower than 25–35, which is considered the optimal value for the initial mixture, causing an excess of N per degradable C [5]. Moreover, it is important to consider the optimal value of moisture (50–65%), because water content along with the bulk density can affect the aeration and microbial O_2 supply [14]. Even in this case, the pig slurry, anaerobic digestate and sewage sludge were the organic materials that showed the highest values of moisture (>90%). This suggests that the adequate selection of bulking agents for both objectives is important to increase the degradable C and improve the physical characteristics of the starting mixtures.

To evaluate the effect of using different raw materials and bulking agents on the evolution of the composting process, temperature parameters were assessed (Figure 1, Table 3). The highest temperatures values (>55 °C) were observed in the mixtures where wood chips were used as a bulking agent (mixtures 1, 2, 3, 10, 11 and 12), and when cereal straw was added as absorbent material (mixtures 2, 3, 11 and 12). These mixtures were also characterized by the highest cumulative DH, suggesting the correct development and intensity of the composting processes, especially when about 40% of wood chips were used. During the active phase of composting, a temperature of 52–60 °C is considered the most favorable, during which mesophilic and thermophilic microorganisms firstly degrade sugars, amino acids and proteins (the mesophilic phase), and afterwards, fats, cellulose, hemicellulose and lignin (the thermophilic phase) [5,28].

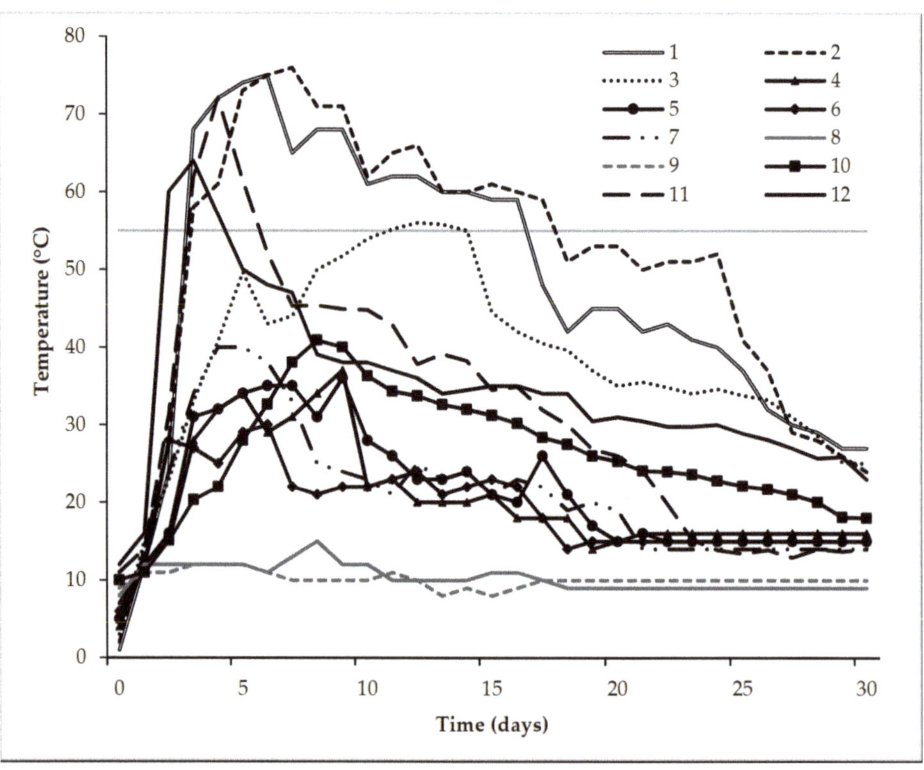

Figure 1. Temperature profiles during the active phase of composting. Horizontal line indicates 55 °C (1–2: [21]; 10–11: [3]; 12: [3]).

Table 3. Temperature parameters during composting.

Mixture	T Max	h > 55 °C	Cumulative DH
1	72	312	36,048
2	74	336	38,160
3	56	96	28,855
4	37	0	15,288
5	36	0	16,128
6	30	0	14,400
7	40	0	15,936
8	12	0	7704
9	12	0	7536
10	41	0	19,586
11	67	96	23,614
12	64	72	26,369

T max: Maximum temperature reached in the active phase; h > 55 °C: Time when the mixture temperature was higher than 55 °C; Cumulative DH: Cumulative degree hours in the active phase.

The other mixtures (from 4 to 9) were prepared with the conifer and broadleaf tree pruning, before and after being chopped, to investigate the effect of different particle sizes on the composting process. In all these mixtures, temperature values >55 °C were not achieved, and this was particularly evident in mixtures 8 and 9 where the olive mill wastes were treated with both types of chipped tree pruning. The high amount of olive mill wastes used in the mixtures resulted in an unbalanced C/N ratio with an excess of C. With this regard, Gigliotti et al. observed a low degradation of organic matter during the composting of husks, probably due to the presence of relatively stable compounds (lipids, polyphenols and pectin) [25]. Conversely, when the pig slurry was co-composted with the tree pruning, the mixture achieved a higher temperature with respect to the olive mill wastes, with an average maximum value of 36 °C, probably due to the chemical composition of pig slurry, as discussed in the following paragraphs.

3.2. Physico-Chemical Parameters of the Mixtures and Temperature Evolution

The mixtures were analyzed for their main physico-chemical characteristics and results are reported in Table 4. Statistical analysis showed that no significant linear relationship existed between the physico-chemical parameters analyzed and the temperature parameters reported in Table 3.

Table 4. Physico-chemical characteristics of the mixtures.

Mixture	Moisture (%)	Volatile Solids (%)	pH	EC (dS m^{-1})	AFP (%)
1 [a]	67 ± 1	73.5 ± 0.1	3.3 ± 0.0	3.3 ± 0.1	53.7
2 [a]	55 ± 1	65.2 ± 0.0	7.3 ± 0.1	1.3 ± 0.1	53.7
3	57 ± 0	86.3 ± 0.0	6.8 ± 0.1	2.5 ± 0.2	44.7
4	57 ± 0	71.0 ± 0.3	8.4 ± 0.1	0.7 ± 0.0	70.8
5	62 ± 0	79.5 ± 0.8	8.3 ± 0.1	0.7 ± 0.0	67.2
6	70 ± 2	72.4 ± 0.7	8.3 ± 0.2	0.8 ± 0.1	56.4
7	64 ± 1	76.8 ± 0.3	8.2 ± 0.1	0.9 ± 0.0	60.9
8	60 ± 1	86.9 ± 0.4	5.9 ± 0.0	2.4 ± 0.2	34.8
9	60 ± 1	89.1 ± 1.2	6.1 ± 0.0	2.1 ± 0.1	46.5
10 [b]	68 ± 1	77.6 ± 0.8	8.2 ± 0.0	2.1 ± 0.1	47.4
11 [b]	62 ± 0	82.1 ± 0.2	7.9 ± 0.1	1.2 ± 0.1	56.4
12 [c]	65 ± 0	69.1 ± 0.3	8.2 ± 0.1	1.5 ± 0.0	51.9

[a] [21]; [b] [3]; [c] [20]; Mean value ± SEM, n = 3; EC: Electrical conductivity, AFP: Air-filled porosity. Data are expressed on dry basis except for pH and EC.

The moisture values ranged from a minimum of 55 to a maximum of 68% in the mixtures where pig slurry (mixture 2) and sewage sludge were processed (mixture 10), respectively, values close to the optimal range (50–65%). This difference was attributed to a large amount of sewage sludge (70%) treated in mixture 10, and also probably to the lack of absorbent materials used, such as the cereal straw. The increase in temperature is related to a variety of factors (e.g., the percentage of porosity, as described by AFP). AFP should be in the range of 35–50% to ensure optimal air circulation [24]. Although mixture 2 showed a slightly higher value of AFP with respect to the optimal range, the correct temperature evolution was evidently not influenced (Table 3, Figure 1). These results suggest that the behavior of the active phase is not strictly dependent on the moisture and porosity of the mixture, especially if both parameters are close to the optimal values. Concerning the volatile solids, the lowest value was observed for mixture 2, suggesting that this parameter did not affect the temperature behavior during the thermophilic phase. This evidence can be supported by the results obtained for mixtures 8 and 9 that, although showing a high content of volatile solids, did not exceed the ambient temperature during the active phase. The analysis of the degradable fraction of organic matter, in terms of C and N content, might be a more suitable parameter to predict the behavior of the thermophilic phase. In a similar experiment in which husks were composted, Gigliotti et al. observed the correct behavior of the active phase to be associated with the biodegradation of labile organic compounds of organic matter, i.e., the water-soluble organic matter [25]. This fraction is easily available for microorganisms and measured in terms of WEOC and was then proposed as a good indicator of the evolution of the process [15,19,25]. This aspect is of particular relevance especially for choosing the bulking agents, which besides having the role of optimizing the physical structure of the mixture, in addition, they should also have a high concentration of low degradable fiber. In this study, the use of tree pruning did not improve the structure of mixtures, as observed by the highest AFP values when these were not chopped and co-composted with pig slurry (mixtures 4 and 5). The attempt to use the chopped tree pruning was useful to decrease the AFP in pig slurry mixtures (mixtures 6 and 7), but the effect of reducing the particle size was more evident when olive mill wastes were treated (mixtures 8 and 9). Hence, it can be stated that the use of chopped tree pruning improves aeration, which represents a key factor for composting [5]. Even if the physical structure was optimized, mixtures 6, 7, 8 and 9 did not achieve the optimal values of temperatures for the active phase of composting. This suggests, once again, that bulk density and AFP are not the only parameters that affect the increase in temperature, but rather the chemical composition of the bulking agents also plays an important role.

With regards to the pH (Table 4), the values ranged from 3.3. to 8.4 (for mixtures 1 and 4, respectively) and these values seem to be outside the optimal range (6.5 to 8.0) for the initial stage of composting, even if a larger pH range may be allowed for ensuring the correct evolution of the process [4], as also demonstrated by our data (Figure 1, Table 4). The results of EC on the starting mixtures showed values in a range of 0.7–3.3 (for mixtures 5 and 1, respectively). By observing the EC results and the temperature profile (Figure 1) of both mixtures 5 and 1, it is possible to also suggest that this parameter did not affect the temperature evolution. It can be useful to quantify the level of soluble salts (EC determination) in order to evaluate possible phytotoxic effects on seed germination or activity of roots using the compost in the agronomic substrates [29].

3.3. Total and Soluble C and N and Temperature Evolution

As previously discussed, physico-chemical parameters of the mixtures were not significantly related to the correct progress of composting. Therefore, other parameters of the mixtures were assessed in this work and related to the temperature profiles during composting. Results of the determination of total organic C (TOC), total Kjeldahl N (TKN), ammonium-N, water-extractable organic C (WEOC), water-extractable N (WEN), water-extractable organic N (WEON) and their ratios were reported in Table 5.

Both TOC and TKN varied in large ranges (from 26.1 to 44.1% and from 0.7 to 2.1% for TOC and TKN, respectively). The large variability of these two parameters was mainly related to the use of different organic wastes. For instance, mixtures 8 and 9 showed the largest amount of TOC and the lowest amount of TKN, and it was mainly related to the large use of olive mill waste in the mixtures, which is known to be rich in organic C and poor in N [30]. Mixture 10, which was mainly composed of sewage sludge (Table 1), showed the lowest TOC and the highest TKN within the tested mixtures, values similar to the composting mixture analyzed in Şevik et al. [31]. Due to the large variability of TOC and TKN, the C/N ratio of the studied mixtures also varied from a minimum value of 14.1 (mixture 10) to a maximum value of 61.7 (mixture 8), and only two mixtures (6 and 7) showed a C/N ratio within the range recommended by Reyes-Torres et al. (25–30) [32].

Ammonium-N content in the mixtures varied according to their composition, and it was higher in the mixtures containing the organic fraction of municipal solid wastes and pig slurry with respect to the mixtures containing olive mill wastes and sewage sludge. Indeed, these last two organic wastes are usually characterized by a low concentration of ammonium-N [33]. Consequently, the TOC/ammonium-N ratio ranged from a maximum value of 882 (mixture 9) to a minimum value of 167 (mixture 1).

Table 5. Total organic C, total Kjeldahl N, ammonium-N, water-extractable C, N and organic N and their ratios in the mixtures.

Parameters							Mixtures					
	1	2	3	4	5	6	7	8	9	10	11	12
TOC (%)	33.4 ± 0.1	26.1 ± 0.1	40.5 ± 0.2	28.4 ± 0.1	31.8 ± 0.0	34.2 ± 0.0	35.9 ± 0.2	43.2 ± 0.4	44.1 ± 0.3	29.6 ± 0.6	31.2 ± 0.2	36.3 ± 0.0
TKN (%)	1.7 ± 0.1	1.2 ± 0.1	1.0 ± 0.1	1.6 ± 0.1	1.6 ± 0.0	1.3 ± 0.1	1.3 ± 0.1	0.7 ± 0.1	1.0 ± 0.2	2.1 ± 0.2	1.7 ± 0.2	1.6 ± 0.1
TOC/TKN	19.6	21.7	40.5	17.7	19.9	26.3	27.6	61.7	44.1	14.1	18.4	22.7
Amm-N (%)	0.20 ± 0.02	0.12 ± 0.01	0.10 ± 0.01	0.11 ± 0.01	0.13 ± 0.01	0.10 ± 0.00	0.12 ± 0.01	0.05 ± 0.01	0.05 ± 0.01	0.10 ± 0.02	0.06 ± 0.01	0.11 ± 0.02
TOC/Amm-N	167	217.5	405	258.2	244.6	342	299.2	864	882	296	520	330
WEOC (%)	4.00 ± 0.01	3.18 ± 0.05	4.20 ± 0.12	1.15 ± 0.05	1.24 ± 0.06	1.15 ± 0.05	1.20 ± 0.06	10.67 ± 0.24	9.36 ± 0.18	2.18 ± 0.10	2.38 ± 0.13	3.54 ± 0.04
WEOC/TKN	2.35	2.65	4.2	0.72	0.78	0.88	0.92	15.24	9.36	1.04	1.4	2.21
WEOC/Amm-N	20	26.5	42	10.5	9.5	11.5	10	213.4	187.2	21.8	39.7	32.2
WEN (%)	0.50 ± 0.04	0.43 ± 0.03	1.00 ± 0.06	0.17 ± 0.01	0.18 ± 0.01	0.18 ± 0.02	0.17 ± 0.00	0.09 ± 0.02	0.08 ± 0.01	0.25 ± 0.02	0.40 ± 0.04	0.52 ± 0.00
WEON (%)	0.30 ± 0.01	0.31 ± 0.01	0.90 ± 0.01	0.06 ± 0.00	0.05 ± 0.00	0.08 ± 0.01	0.05 ± 0.00	0.04 ± 0.01	0.03 ± 0.00	0.15 ± 0.02	0.34 ± 0.01	0.41 ± 0.02
TOC/WEN	66.8	60.7	40.5	167.1	176.7	190	211.2	480	551.3	118.4	78	69.8
WEOC/WEN	8	7.4	4.2	6.8	6.9	6.4	7.1	118.6	117	8.7	5.9	6.8
TOC/WEON	111.3	84.2	45	473.3	636	427.5	718	1080	1470	197.3	91.8	88.5
WEOC/WEON	13.3	10.3	4.7	19.2	24.8	14.4	24	266.8	312	14.5	7.0	8.6

Mean value ± SEM, n = 3; TOC: Total organic C, TKN: Total Kjeldahl N, Amm-N: Ammonium-N, WEOC: Water-extractable organic C, WEN: Water-extractable N, WEON: Water-extractable organic N. Data are expressed on dry basis.

WEOC and WEN were determined on mixtures' samples to evaluate the relationship between the soluble forms of these two elements and the evolution of composting. WEOC showed large variability between the mixtures analyzed, mainly depending on their composition. The highest values were detected for mixtures 8 and 9 (10.7 and 9.4%, respectively) and this was expected since olive mill wastes are usually rich in soluble organic matter. For instance, Gianico et al. reported that 87% of the chemical oxygen demand of olive milling residues was represented by soluble molecules [34]. Mixtures containing olive mill wastes (8 and 9) were characterized by the highest WEOC/TKN (15.2 and 9.4) and WEOC/ammonium-N (213 and 187) ratios due to their high content of soluble organic matter and low content of N. The largest amounts of WEN were detected in the mixtures containing anaerobically digested organic wastes (mixture 3 and 12). This was expected since it is well known that during anaerobic digestion, organic N is mineralized [35], leading to the formation of ammonium-N, which increases the amount of water-extractable N. With regard to WEON, mixtures 4, 5, 6, 7, 8 and 9 showed a content of WEON ten times lower than the other mixtures, and this was mainly related to the fact that even if most of the mixtures were characterized by an acceptable level of TKN (mixtures 4 to 7), it was present mainly in non-soluble forms. When the ratios between the analyzed parameters were determined (TOC/WEN, WEOC/WEN, TOC/WEON and WEOC/WEON), it was interesting to observe that the highest values of these parameters were always presented by mixtures 8 and 9 (olive mill waste mixtures). This was a consequence of the large amount of organic matter (both in total and soluble forms) and the low amount of N in the olive mill wastes. Similarly, mixtures 4, 5, 6 and 7 (pig slurry mixtures) also showed high values of TOC/WEN and TOC/WEON ratios (ranging from 167 and 211 and from 428 to 718, respectively).

The relationship between mixtures' chemical parameters shown in Table 5 and the correct development of composting in terms of temperature (Figure 1, Table 3) was then evaluated, and the results are reported in Table 6. Although the parameter h > 55 °C did not show significant correlations with the analyzed parameters, the maximum temperature and cumulative DH were found to be significantly correlated with some of the above-described parameters. The strongest and significant correlations were found between temperature parameters and WEN, WEON, TOC/WEN, TOC/ammonium-N, TOC/WEON, WEOC/WEN and WEOC/WEON ratios. Interestingly, WEN and WEON showed a positive correlation with the correct development of the active phase of composting, meaning that the presence of soluble N form positively affects the composting. Conversely, TKN was not significantly correlated with the temperature parameters. This is in accordance with Puyuelo et al. and Trémier et al. who reported that, more so than the total amount of N, it is the soluble forms of the element that should be taken into consideration to optimize composting [14,16]. This is because soluble N compounds (i.e., amino acids, small peptides and ammonium-N) are easily available forms of N that microorganisms can use for their metabolism and growth [36,37]. With respect to TOC/WEN, TOC/ammonium-N, TOC/WEON, WEOC/WEN and WEOC/WEON ratios and their relationship with temperature parameters during the active phase of composting, the correlation found in the present work was negative. This means that an elevated ratio between organic C (both in its total and soluble forms) and water-soluble N forms negatively affect the development of composting. This allows one to affirm that soluble N represents the limiting factor for the correct development of composting, with particular regard to the active phase. The TOC/TKN ratio, which is the most-used parameter for the formulation of a composting mixture, does not take into account the fact that C and N may be present in the organic wastes in recalcitrant forms that cannot be quickly degraded and mineralized by composting microorganisms. For instance, mixtures 4 to 7 in this work showed values of TOC/TKN compatible with composting, taking into account commonly used values [10,32]. Nevertheless, none of these mixtures reached 55 °C during the active phase of composting and the cumulative DH was significantly lower than the other mixtures. This can be attributable to the fact that the main components of these mixtures (pig slurry and tree pruning) are characterized by low contents of easily degradable forms of C and N [38]. The mixtures that showed the worst temperature parameters during the active phase of composting (mixture 8 and 9) were characterized by high values of TOC/WEN, WEOC/WEN, TOC/WEON and WEOC/WEON, suggesting that the presence of a large amount of organic matter, even in soluble forms, is not a sufficient parameter to predict the correct evolution of temperature.

Table 6. Linear regression between chemical parameters of the mixtures and temperature evolution during the active phase of composting.

x	T Max (y)				h > 55 °C (y)				Cumulative DH (y)			
	m	q	R^2	r	m	q	R^2	r	m	q	R^2	r
TOC	−2.1080	117.93	0.3124	n.s.	−7.8552	347.46	0.1351	n.s.	−853.54	50299	0.2379	n.s.
TKN	24.3370	11.01	0.1919	n.s.	18.7950	49.68	0.0036	n.s.	7560	10217	0.0860	n.s.
TOC/TKN	−0.9250	70.85	0.3588	0.5990 *	−2.1832	136.84	0.0623	n.s.	−336.34	30174	0.2204	n.s.
Amm-N	297.18	14.12	0.3152	n.s.	1694.8	−100.54	0.3193	n.s.	155288	4626	0.3998	0.6323 *
TOC/Amm-N	−0.0596	69.03	0.4314	0.6568 *	−0.2070	159.25	0.1623	n.s.	−26.81	31584	0.4060	0.6372 *
WEOC	−3.1730	56.78	0.2173	n.s.	−1.4462	81.33	0.0014	n.s.	−1044.3	24653	0.1093	n.s.
WEOC/TKN	−2.7408	54.62	0.3159	n.s.	−3.9539	89.75	0.0205	n.s.	−978.41	24207	0.1870	n.s.
WEOC/Amm-N	−0.1926	55.10	0.3945	0.6281 *	−0.4063	97.13	0.0547	n.s.	−75.89	24750	0.2845	n.s.
WEN	54.7250	26.06	0.4417	0.6646 *	205.60	4.55	0.1942	n.s.	27274	11324	0.5095	0.7138 **
WEON	50.6530	32.75	0.3551	0.5959 *	175.39	33.32	0.1326	n.s.	25059	14704	0.4036	0.6353 *
TOC/WEN	−0.1101	64.87	0.7358	0.8578 **	−0.3418	137.45	0.2209	n.s.	−47.13	29276	0.6263	0.7914 **
TOC/WEON	−0.0404	62.94	0.7465	0.8640 **	−0.1367	136.48	0.2667	n.s.	−17.58	28580	0.6577	0.8110 **
WEOC/WEN	−0.2100	54.31	0.5068	0.7119 **	−0.7904	95.75	0.0785	n.s.	−141.06	24327	0.3729	0.6107 *
WEOC/WEON	−0.1467	53.78	0.5403	0.7351 **	−0.3477	96.60	0.0945	n.s.	−58.98	24298	0.4054	0.6367 *

n.s.: Not significant, *: Significant at $p < 0.05$, ** significant at $p < 0.01$, n = 12; T max: Maximum temperature during the active phase, DH: Degree hours, TOC: Total organic C, TKN: Total Kjeldahl N, Amm-N: Ammonium-N, WEOC: Water-extractable organic C, WEN: Water-extractable N, WEON: Water-extractable organic N.

Based on the results reported in Tables 5 and 6, an attempt to propose the optimal values of chemical parameters to optimize the active phase of composting was carried out. Nevertheless, it should be highlighted that these are only preliminary results that need experimental confirmation. If the WEN is considered, its concentration should be higher than 0.4% (*w/w*) in order to ensure an optimal amount of easily available N. Indeed, in the present work, all the mixtures that showed the correct development of the active phase were characterized by a WEN concentration higher than 0.4% (*w/w*). Within the ratios studied, TOC/WEN and TOC/WEON appeared to be the most promising parameters that should be taken into consideration during the preparation of composting mixtures, since they were found to correlate significantly with the correct development of the process. With regard to the results reported in the present work, the TOC/WEN ratio should range between 40 and 80 to achieve high temperatures and cumulative DH, whereas the TOC/WEON ratio should range between 40 and 120. The proposed values should ensure a balanced ratio between C and N amounts and their availability, allowing for the quick growth of microorganisms and correct behavior of composting.

3.4. Mixture Characteristics and Their Effect on Compost Quality

The main characteristics of the compost produced from the 12 mixtures studied and the limit values for quality compost production established by Italian legislation are reported in Table 7.

TOC and TKN concentrations in the composts studied varied depending on the initial composition of the mixtures. The highest amount of TOC was found in the compost produced from olive mill wastes (mixtures 8 and 9), whereas the highest amount of TKN was found in the compost produced from sewage sludge and anaerobically digested pharmaceutical wastewater (mixtures 10, 11 and 12).

OM-loss values higher than 30–40% are usually considered an index of a correctly developed composting process [39,40]. In the present work, only five composts showed optimal values of OM-loss (composts from mixtures 1, 2, 3, 11 and 12), whereas compost from mixture 10 showed a value near to the limit (30.9%). In the other composts (4, 5, 6, 7, 8 and 9), OM-loss ranged from 5.8 to 18.2%, demonstrating that the lack of an increase in temperature in the active phase compromised the whole composting process, also negatively affecting the decrease in moisture. Indeed, most of the composts produced from mixtures that did not undergo a proper increase in temperature during the active phase (composts from mixtures 4, 6, 8, 9 and 10) showed high moisture values (above 50% w/w). Conversely, when an adequate temperature profile was achieved during composting, heating caused a significant decrease in moisture, as expected at the end of composting [41].

The pH of composts showed values expected at the end of composting. Indeed, compost usually shows a sub-alkaline pH due to the mineralization of volatile fatty acids and the release of ammonium-N after proteins' hydrolysis [1,2]. Soluble salts' content (estimated from the electrical conductivity, EC) in the composts varied in relation to the composition of the initial mixtures and ranged from 0.6 dS m^{-1} to 4.6 dS m^{-1} (composts obtained from mixtures 9 and 10, respectively).

The compost maturity was evaluated by the determination of the content of humic and fulvic acids (HA + FA), as well as the germination index (GI). Only compost produced by the mixtures containing olive mill wastes (mixtures 8 and 9) showed low values of HA + FA and GI (6.8 and 6.3% of HA + FA, 58.7 and 54.3% of GI). Considering that these were the two mixtures showing the worst temperature profiles, these results confirmed that a proper active phase of composting is also essential for the correct behavior of the curing phase. Indeed, the maturation of compost mainly occurs during the curing phase, when the phytotoxic compounds are mineralized and the organic matter is stabilized [42].

The comparison of the results of compost characterization with the Italian limits for quality compost production showed that five of the seven composts produced by mixtures that did not show a proper temperature profile in the active phase did not comply with the legal requirements for compost commercialization (compost from mixtures 4, 6, 8, 9 and 10) [43]. Whereas composts from mixtures 4, 6 and 10 showed excessive moisture, composts from mixtures 8 and 9 (olive mill waste mixtures) did not comply with four parameters (maximum C/N and moisture, minimum content of HA + FA and minimum GI value). Considering these results, ensuring the correct development of the active phase of composting appears mandatory to obtain a high quality of the compost produced. Given that composting is an energy-consuming process, the importance of optimizing the composition of initial mixtures is clear, as is using appropriate calculation tools such as the one described in Calisti et al. [44]. If composting mixtures are not well optimized, the final compost might not comply with legal limits and should then be disposed of by incineration or landfilling, causing environmental and economic issues [45,46].

The possible relationship between the new parameters proposed to optimize composting mixtures (WEN, WEON, TOC/WEN, TOC/WEON, WEOC/WEN and WEOC/WEON) and the main physico-chemical characteristic of composts was investigated by correlation analysis and the results are reported in Table 8.

Interestingly, the parameters of initial mixtures that were correlated with the temperature parameters in the active phase of composting were also correlated to most of the selected physico-chemical characteristics of composts. Obviously, TOC and C/N of composts were positively correlated to TOC/WEN, TOC/WEON, WEOC/WEN and WEOC/WEON ratios, and the EC of compost was positively correlated to WEN and WEON in the mixtures. This latter correlation was expected since soluble N compounds (i.e., ammonium-N) increase the electrical conductivity of the water-compost extract. In this context, it must be taken into account that an excessive WEN or WEON concentration in the initial mixture can result in high EC in the composts, with potential phytotoxic effects [47].

Table 7. Main characteristics of the compost produced from the mixtures studied and limit values for quality compost production [43].

Parameters							Mixtures						
	1 [a]	2 [a]	3	4	5	6	7	8	9	10 [b]	11 [b]	12 [c]	Limit Value [d]
TOC (%)	20.2 ± 0.8	21.7 ± 0.2	21.6 ± 0.2	25.4 ± 0.1	24.8 ± 0.0	23.7 ± 0.1	25.2 ± 0.0	36.2 ± 0.0	36.2 ± 0.5	30.7 ± 0.6	23.2 ± 0.3	30.4 ± 0.2	20 (min)
TKN (%)	2.2 ± 0.1	2.1 ± 0.1	1.0 ± 0.2	1.4 ± 0.1	1.3 ± 0.1	1.2 ± 0.2	1.3 ± 0.1	1.2 ± 0.0	1.3 ± 0.0	3.2 ± 0.2	2.5 ± 0.2	2.7 ± 0.2	-
TOC/TKN	9.2	10.3	21.6	18.1	19.1	19.8	19.4	30.2	27.8	9.6	9.3	11.3	25
OM-loss (%)	62.1 ± 0.3	80.2 ± 0.8	66.0 ± 0.4	17.0 ± 0.1	17.5 ± 0.0	18.2 ± 0.2	17.9 ± 0.1	6.2 ± 0.6	5.8 ± 0.5	30.9 ± 0.4	63.1 ± 1.3	38.8 ± 1.0	-
Moisture (%)	36.9 ± 0.2	31.5 ± 0.6	44.9 ± 0.5	53.7 ± 0.6	42.8 ± 0.8	55.9 ± 0.5	48.1 ± 0.0	59.2 ± 0.4	63.6 ± 0.7	63.0 ± 0.6	48.3 ± 0.4	48.1 ± 0.3	50
pH	7.3 ± 0.1	7.2 ± 0.0	7.9 ± 0.0	8.0 ± 0.1	7.8 ± 0.1	8.0 ± 0.1	7.9 ± 0.0	7.0 ± 0.0	7.2 ± 0.0	7.9 ± 0.0	8.3 ± 0.2	8.4 ± 0.1	6–8.5
EC (dS m^{-1})	2.3 ± 0.0	2.1 ± 0.0	3.7 ± 0.2	1.4 ± 0.2	1.4 ± 0.1	1.5 ± 0.1	1.4 ± 0.2	0.8 ± 0.1	0.6 ± 0.0	4.6 ± 0.0	1.9 ± 0.2	3.5 ± 0.2	-
HA + FA (%)	8.4 ± 0.1	9.3 ± 0.1	11.4 ± 0.2	7.7 ± 0.2	8.3 ± 0.1	8.2 ± 0.2	8.4 ± 0.2	6.8 ± 0.1	6.3 ± 0.0	8.3 ± 0.1	11.7 ± 0.1	12.1 ± 0.1	7 (min)
GI (%)	112.9 ± 1.4	103.6 ± 0.8	100.6 ± 2.0	72.9 ± 2.3	69.3 ± 1.6	70.8 ± 1.0	69.4 ± 0.5	58.7 ± 0.8	54.3 ± 1.0	78.3 ± 1.4	92.3 ± 2.3	83.7 ± 0.2	60 (min)

[a] [21]; [b] [3]; [c] [20]; [d] [43]; Mean value ± SEM, n = 3; TOC: Total organic C, TKN: Total Kjeldahl N, OM: Organic matter, EC: Electrical conductivity, HA + FA: Humic and fulvic acids, GI: Germination index. Data are expressed on dry basis except for pH and EC.

Table 8. Pearson's correlation coefficients between selected parameters of the mixtures and composts studied.

		Mixture						Compost							
		WEN	WEON	TOC/WEN	TOC/WEON	WEOC/WEN	WEOC/WEON	TOC	TKN	C/N	OM-loss	Moisture	pH	EC	HA + FA	GI
Mixture	WEN	1														
	WEON	0.9880 **	1													
	TOC/WEN	−0.6910 *	−0.6199 *	1												
	TOC/WEON	−0.6798 *	−0.6798 *	0.9725 **	1											
	WEOC/WEN	n.s.	n.s.	0.9344 **	0.8438 **	1										
	WEOC/WEON	n.s.	n.s.	0.9538 **	0.8820 **	0.9940 **	1									
Compost	TOC	n.s.	n.s.	0.7624 **	0.6919 *	0.7904 **	0.7884 **	1								
	TKN	n.s.	n.s.	n.s.	n.s.	n.s.	n.s.	n.s.	1							
	C/N	n.s.	n.s.	0.8397 **	0.8379 **	0.7524 **	0.7647 **	0.5830 *	−0.8119 **	1						
	OM-loss	0.7599 **	0.7243 **	−0.7374 **	−0.7863 **	n.s.	n.s.	−0.6904 *	n.s.	−0.6836 *	1					
	Moisture	n.s.	n.s.	n.s.	n.s.	n.s.	n.s.	0.5823 *	n.s.	0.5777 *	n.s.	1				
	pH	n.s.	n.s.	n.s.	n.s.	−0.6650 *	−0.6507 *	n.s.	n.s.	n.s.	n.s.	n.s.	1			
	EC	0.7798 **	0.7643 **	−0.7161 **	−0.7575 **	n.s.	n.s.	n.s.	0.6419 *	−0.5848 *	n.s.	n.s.	n.s	1		
	HA + FA	0.7351 **	0.7450 **	−0.7222 **	−0.7501 **	−0.5927 *	−0.6198 *	n.s.	n.s.	n.s.	0.6774 *	n.s.	0.6722 *	0.5898 *	1	
	GI	0.7626 **	0.6892 *	−0.7984 **	−0.8320 **	−0.6105 *	−0.6418 *	−0.7870 **	n.s.	−0.7343 **	0.9476 **	n.s.	n.s.	n.s.	0.6065 *	1

n.s.: not significant, *: significant at $p < 0.05$, **: significant at $p < 0.01$, n = 12. TOC: total organic C, TKN: total Kjeldahl N, WEOC: water extractable organic C, WEN: water extractable N, WEON: water extractable organic N, OM: organic matter, EC: electrical conductivity, HA + FA: humic and fulvic acids, GI: germination index.

OM-loss was positively correlated with WEN and WEON, demonstrating that soluble N compounds are essential to sustain microbial metabolism and, thus, composting effectiveness. Conversely, OM loss was negatively correlated with TOC/WEN and TOC/WEON ratios, confirming that high values of these ratios cannot ensure a balanced substrate for microbial growth during composting.

Moreover, the parameters commonly used to evaluate the maturity were correlated with the new parameters proposed to optimize composting mixtures. Both HA + FA and GI were positively correlated with WEN and WEON, and negatively correlated with TOC/WEN, TOC/WEON, WEOC/WEN and WEOC/WEON ratios. Considering these results, the parameters proposed to optimize composting mixtures also appear suitable to ensure the production of good-quality composts, in terms of hygenization, stabilization and maturation.

4. Conclusions

The present work aimed to assess new parameters for composting in order to better explain the process and to optimize it, in terms of correct temperature evolution and high-quality compost. Although physico-chemical (i.e., moisture, air-filled porosity) and chemical (i.e., TOC/TKN) parameters were commonly used to optimize composting mixtures, they did not ensure the correct development of the aerobic process. Conversely, a significant positive correlation was found between temperature evolution during the active phase of composting and soluble forms of N (WEN and WEON) in the starting mixtures, demonstrating that the easily available N compounds play a key role in the correct development of composting. On the contrary, high TOC/WEN, TOC/WEON, WEOC/WEN and WEOC/WEON ratios resulted in a low temperature and poor quality of the final compost.

Based on these results, it was suggested that the study of soluble forms of C and N may help to predict the mineralization rate during the active phase and the correct increase in temperature. In particular, WEN and the TOC/WEN ratio might be useful parameters to evaluate the aptitude of a starting mixture to be composted. Nevertheless, future research is needed to confirm the results reported in this work, by studying mixtures with fixed values of moisture, AFP, volatile solids, pH and C/N, in order to better understand the role of WEN and TOC/WEN ratio parameters in the degradation of organic matter during composting.

Author Contributions: Conceptualization: M.C. and D.P.; methodology: M.C. and D.P.; investigation: M.C. and D.P.; data curation: M.C.; writing—original draft preparation: M.C. and D.P.; writing—review and editing: G.G., P.P., R.C. and L.R.; supervision: G.G. and P.P. All authors have read and agreed to the published version of the manuscript.

Funding: This research was funded by the EU Rural Development Plan 2014–2020 of the Umbria Region, Italy (Action 16.1; grant number 84250135484).

Institutional Review Board Statement: Not applicable.

Informed Consent Statement: Not applicable.

Data Availability Statement: The data presented in this study are available in the study itself.

Acknowledgments: The authors acknowledge Gina Pero for laboratory analysis.

Conflicts of Interest: The authors declare no conflict of interest.

References

1. Cerda, A.; Artola, A.; Font, X.; Barrena, R.; Gea, T.; Sánchez, A. Composting of food wastes: Status and challenges. *Bioresour. Technol.* **2018**, *248*, 57–67. [CrossRef]
2. Wang, S.; Zeng, Y. Ammonia emission mitigation in food waste composting: A review. *Bioresour. Technol.* **2018**, *248*, 13–19. [CrossRef]
3. Cucina, M.; Tacconi, C.; Sordi, S.; Pezzolla, D.; Gigliotti, G.; Zadra, C. Valorization of a pharmaceutical organic sludge through different composting treatments. *Waste Manag.* **2018**, *74*, 203–212. [CrossRef]

4. Proietti, P.; Calisti, R.; Gigliotti, G.; Nasini, L.; Regni, L.; Marchini, A. Composting optimization: Integrating cost analysis with the physical-chemical properties of materials to be composted. *J. Clean. Prod.* **2016**, *137*, 1086–1099. [CrossRef]

5. Bernal, M.P.; Alburquerque, J.A.; Moral, R. Composting of animal manures and chemical criteria for compost maturity assessment. A review. *Bioresour. Technol.* **2009**, *100*, 5444–5453. [CrossRef] [PubMed]

6. Lin, L.; Xu, F.; Ge, X.; Li, Y. Improving the sustainability of organic waste management practices in the food-energy-water nexus: A comparative review of anaerobic digestion and composting. *Renew. Sustain. Energy Rev.* **2018**, *89*, 151–167. [CrossRef]

7. Dhamodharan, K.; Varma, V.S.; Veluchamy, C.; Pugazhendhi, A.; Rajendran, K. Emission of volatile organic compounds from composting: A review on assessment, treatment and perspectives. *Sci. Tot. Environ.* **2019**, *695*, 133725. [CrossRef] [PubMed]

8. Nasini, L.; De Luca, G.D.; Ricci, A.; Ortolani, F.; Caselli, A.; Massaccesi, L.; Regni, L.; Gigliotti, G.; Proietti, P. Gas emissions during olive mill waste composting under static pile conditions. *Int. Biodet. Biodeg.* **2016**, *107*, 70–76. [CrossRef]

9. Cáceres, R.; Coromina, N.; Malińska, K.; Marfà, O. Evolution of process control parameters during extended co-composting of green waste and solid fraction of cattle slurry to obtain growing media. *Bioresour. Technol.* **2015**, *179*, 398–406. [CrossRef]

10. Sharma, D.; Yadav, K.D.; Kumar, S. Biotransformation of flower waste composting: Optimization of waste combinations using response surface methodology. *Bioresour. Technol.* **2018**, *270*, 198–207. [CrossRef]

11. Adhikari, B.K.; Barrington, S.; Martinez, J.; King, S. Effectiveness of three bulking agents for food waste composting. *Waste Manag.* **2009**, *29*, 197–203. [CrossRef]

12. Barrena, R.; Turet, J.; Busquets, A.; Farrés, M.; Font, X.; Sánchez, A. Respirometric screening of several types of manure and mixtures intended for composting. *Bioresour. Technol.* **2011**, *102*, 1367–1377. [CrossRef]

13. Bueno, P.; Yanez, R.; Rivera, A.; Díaz, M.J. Modelling of parameters for optimization of maturity in composting trimming residues. *Bioresour. Technol.* **2009**, *100*, 5859–5864. [CrossRef]

14. Trémier, A.; Teglia, C.; Barrington, S. Effect of initial physical characteristics on sludge compost performance. *Bioresour. Technol.* **2009**, *100*, 3751–3758. [CrossRef]

15. Paradelo, R.; Moldes, A.B.; Barral, M.T. Evolution of organic matter during the mesophilic composting of lignocellulosic winery wastes. *J. Environ. Manag.* **2013**, *116*, 18–26. [CrossRef]

16. Puyuelo, B.; Ponsá, S.; Gea, T.; Sánchez, A. Determining C/N ratios for typical organic wastes using biodegradable fractions. *Chemosphere* **2011**, *85*, 653–659. [CrossRef]

17. Samudrika, K.P.D.; Ariyawansha, R.T.K.; Basnayake, B.F.A.; Siriwardana, A.N. Optimization of biochar additions for enriching nitrogen in active phase low-temperature composting. *Org. Agric.* **2020**, *10*, 449–463. [CrossRef]

18. Tambone, F.; Terruzzi, L.; Scaglia, B.; Adani, F. Composting of the solid fraction of digestate derived from pig slurry: Biological processes and compost properties. *Waste Manag.* **2015**, *35*, 55–61. [CrossRef] [PubMed]

19. Said-Pullicino, D.; Erriquens, F.G.; Gigliotti, G. Changes in the chemical characteristics of water-extractable organic matter during composting and their influence on compost stability and maturity. *Bioresour. Technol.* **2007**, *98*, 1822–1831. [CrossRef]

20. Cucina, M.; Zadra, C.; Marcotullio, M.C.; Di Maria, F.; Sordi, S.; Curini, M.; Gigliotti, G. Recovery of energy and plant nutrients from a pharmaceutical organic waste derived from a fermentative biomass: Integration of anaerobic digestion and composting. *J. Environ. Chem. Eng.* **2017**, *5*, 3051–3057. [CrossRef]

21. Tacconi, C.; Cucina, M.; Zadra, C.; Gigliotti, G.; Pezzolla, D. Plant nutrients recovery from aflatoxin B1 contaminated corn through co-composting. *J. Environ. Chem. Eng.* **2019**, *7*, 103046. [CrossRef]

22. American Public Health Association (APHA); Water Environment Federation. *Standard Methods for the Examination of Water and Wastewater*; American Public Health Association: Washington, DC, USA, 2016.

23. The US Department of Agriculture and The US Composting Council. *Test. Methods for the Examination of Composting and Compost*; Edaphos International: Houston, TX, USA, 2016.

24. Alburquerque, J.A.; McCartney, D.; Yu, S.; Brown, L.; Leonard, J.J. Air space in composting research: A literature review. *Compost Sci. Util.* **2008**, *16*, 159–170. [CrossRef]

25. Gigliotti, G.; Proietti, P.; Said-Pullicino, D.; Nasini, L.; Pezzolla, D.; Rosati, L.; Porceddu, P.R. Co-composting of olive husks with high moisture contents: Organic matter dynamics and compost quality. *Int. Biodeter. Biodegr.* **2012**, *67*, 8–14. [CrossRef]

26. Ciavatta, C.; Govi, M.; Antisari, L.V.; Sequi, P. Characterization of humified compounds by extraction and fractionation on solid polyvinylpyrrolidone. *J. Chromatogr. A* **1990**, *509*, 141–146. [CrossRef]

27. Agnew, J.M.; Leonard, J.J. The physical properties of compost. *Compost Sci. Util.* **2003**, *11*, 238–264. [CrossRef]

28. Miller, F.C. Composting as a process based on the control of ecologically selective factors. In *Soil Microbial Ecology: Applications in Agricultural and Environmental Management*; Metting., F.B., Ed.; CRC Press: Ottawa, ON, Canada, 1992; pp. 515–544.

29. Arslan, E.I.; Ünlü, A.; Topal, M. Determination of the effect of aeration rate on composting of vegetable–fruit wastes. *CLEAN–Soil Air Water.* **2011**, *39*, 1014–1021. [CrossRef]

30. Nunes, M.A.; Costa, A.S.; Bessada, S.; Santos, J.; Puga, H.; Alves, R.C.; Freitas, V.; Oliveira, M.B. Olive pomace as a valuable source of bioactive compounds: A study regarding its lipid-and water-soluble components. *Sci. Tot. Environ.* **2018**, *644*, 229–236. [CrossRef]

31. Şevik, F.; Tosun, İ.; Ekinci, K. The effect of FAS and C/N ratios on co-composting of sewage sludge, dairy manure and tomato stalks. *Waste Manag.* **2018**, *80*, 450–456. [CrossRef]

32. Reyes-Torres, M.; Oviedo-Ocaña, E.R.; Dominguez, I.; Komilis, D.; Sánchez, A. A systematic review on the composting of green waste: Feedstock quality and optimization strategies. *Waste Manag.* **2018**, *77*, 486–499. [CrossRef]

33. Haddadin, M.S.; Haddadin, J.; Arabiyat, O.I.; Hattar, B. Biological conversion of olive pomace into compost by using *Trichoderma harzianum* and *Phanerochaete chrysosporium*. *Bioresour. Technol.* **2009**, *100*, 4773–4782. [CrossRef]

34. Gianico, A.; Braguglia, C.M.; Mescia, D.; Mininni, G. Ultrasonic and thermal pretreatments to enhance the anaerobic bioconversion of olive husks. *Bioresour. Technol.* **2013**, *147*, 623–626. [CrossRef]

35. Pigoli, A.; Zilio, M.; Tambone, F.; Mazzini, S.; Schepis, M.; Meers, E.; Schoumans, O.; Giordano, A.; Adani, F. Thermophilic anaerobic digestion as suitable bioprocess producing organic and chemical renewable fertilizers: A full-scale approach. *Waste Manag.* **2021**, *124*, 356–367. [CrossRef]

36. Jamroz, E.; Bekier, J.; Medynska-Juraszek, A.; Kaluza-Haladyn, A.; Cwielag-Piasecka, I.; Bednik, M. The contribution of water extractable forms of plant nutrients to evaluate MSW compost maturity: A case study. *Sci. Rep.* **2020**, *10*, 1–9. [CrossRef]

37. Wang, M.; Ma, L.; Kong, Z.; Wang, Q.; Fang, L.; Liu, D.; Shen, Q. Insights on the aerobic biodegradation of agricultural wastes under simulated rapid composting conditions. *J. Clean Prod.* **2019**, *220*, 688–697. [CrossRef]

38. Martín-Mata, J.; Lahoz-Ramos, C.; Bustamante, M.A.; Marhuenda-Egea, F.C.; Moral, R.; Santos, A.; Bernal, M.P. Thermal and spectroscopic analysis of organic matter degradation and humification during composting of pig slurry in different scenarios. *Environ. Sci. Pollut. R.* **2016**, *23*, 17357–17369. [CrossRef] [PubMed]

39. Doublet, J.; Francou, C.; Poitrenaud, M.; Houot, S. Sewage sludge composting: Influence of initial mixtures on organic matter evolution and N availability in the final composts. *Waste Manag.* **2010**, *30*, 1922–1930. [CrossRef] [PubMed]

40. Fornes, F.; Mendoza-Hernández, D.; García-de-la-Fuente, R.; Abad, M.; Belda, R.M. Composting versus vermicomposting: A comparative study of organic matter evolution through straight and combined processes. *Bioresour. Technol.* **2012**, *118*, 296–305. [CrossRef] [PubMed]

41. Thomas, C.; Idler, C.; Ammon, C.; Amon, T. Effects of the C/N ratio and moisture content on the survival of ESBL-producing *Escherichia coli* during chicken manure composting. *Waste Manag.* **2020**, *105*, 110–118. [CrossRef] [PubMed]

42. Wang, K.; Mao, H.; Wang, Z.; Tian, Y. Succession of organics metabolic function of bacterial community in swine manure composting. *J. Hazard. Mater.* **2018**, *360*, 471–480. [CrossRef] [PubMed]

43. Decreto Legislativo 29 Aprile 2010, n. 75. In *Riordino e Revisione della Disciplina in Materia di Fertilizzanti, a Norma dell'Articolo 13 della Legge 7 Luglio 2009 n. 88*; Gazzetta Ufficiale n. 121-Suppl. Ordin. n.106; Governo Italiano: Roma, Italy, 2010.

44. Calisti, R.; Regni, L.; Proietti, P. Compost-recipe: A new calculation model and a novel software tool to make the composting mixture. *J. Clean. Prod.* **2020**, *270*, 122427. [CrossRef]

45. Doña-Grimaldi, V.M.; Palma, A.; Ruiz-Montoya, M.; Morales, E.; Díaz, M.J. Energetic valorization of MSW compost valorization by selecting the maturity conditions. *J. Environ. Manag.* **2019**, *238*, 153–158. [CrossRef]

46. Majdinasab, A.; Zhang, Z.; Yuan, Q. Modelling of landfill gas generation: A review. *Rev. Environ. Sci. Biol.* **2017**, *16*, 361–380. [CrossRef]

47. Siles-Castellano, A.B.; López, M.J.; López-González, J.A.; Suárez-Estrella, F.; Jurado, M.M.; Estrella-González, M.J.; Moreno, J. Comparative analysis of phytotoxicity and compost quality in industrial composting facilities processing different organic wastes. *J. Clean Prod.* **2020**, *252*, 119820. [CrossRef]

Article

Influence of Recycled Waste Compost on Soil Food Webs, Nutrient Cycling and Tree Growth in a Young Almond Orchard

Amanda K. Hodson [1,*], Jordan M. Sayre [2], Maria C. C. P. Lyra [3] and Jorge L. Mazza Rodrigues [2]

[1] Department of Entomology and Nematology, University of California Davis, Davis, CA 95616, USA
[2] Department of Land, Air and Water Resources, University of California Davis, Davis, CA 95616, USA; jmsayre@ucdavis.edu (J.M.S.); jmrodrigues@ucdavis.edu (J.L.M.R.)
[3] Laboratory of Genomics, Agronomic Institute of Pernambuco-IPA, Recife 50761-000, Brazil; mccatanho@gmail.com
* Correspondence: akhodson@ucdavis.edu

Abstract: Composting is an effective strategy to process agricultural and urban waste into forms that may be beneficial to crops. The objectives of this orchard field study were to characterize how a dairy manure compost and a food waste compost influenced: (1) soil nitrogen and carbon pools, (2) bacterial and nematode soil food webs and (3) tree growth and leaf N. The effects of composts were compared with fertilized and unfertilized control plots over two years in a newly planted almond orchard. Both dairy manure compost and food waste compost increased soil organic matter pools, as well as soil nitrate and ammonium at certain time points. Both composts also distinctly altered bacterial communities after application, specifically those groups with carbon degrading potential, and increased populations of bacterial feeding nematodes, although in different timeframes. Unique correlations were observed between nematode and bacterial groups within compost treatments that were not present in controls. Food waste compost increased trunk diameters compared to controls and had greater relative abundance of herbivorous root tip feeding nematodes. Results suggest that recycled waste composts contribute to biologically based nitrogen cycling and can increase tree growth, mainly within the first year after application.

Keywords: organic waste; manure; nematode community; 16S; bacterial community

Citation: Hodson, A.K.; Sayre, J.M.; Lyra, M.C.C.P.; Rodrigues, J.L.M. Influence of Recycled Waste Compost on Soil Food Webs, Nutrient Cycling and Tree Growth in a Young Almond Orchard. *Agronomy* **2021**, *11*, 1745. https://doi.org/10.3390/agronomy11091745

Academic Editors: Luca Regni and Mirko Cucina

Received: 30 July 2021
Accepted: 26 August 2021
Published: 30 August 2021

Publisher's Note: MDPI stays neutral with regard to jurisdictional claims in published maps and institutional affiliations.

1. Introduction

Soil health has been defined as "the capacity of the soil to function as a vital living ecosystem that supports plants, animals, and humans" [1]. It is determined by interactions between microbial communities, soil physical and chemical factors, and management decisions [2], encompassing biological attributes such as biodiversity, food web structure and ecosystem functioning [3]. Increasing soil organic carbon (SOC) serves as the foundation for building healthy soils [4]. As an important indicator of soil quality, SOC enhances crop productivity by improving water holding capacity, aggregation, nutrient transformation and microbial biomass [5]. While intensive agriculture depletes SOC, land management practices that lead to increases in SOC reverse this trend, enhancing productivity and environmental quality [6].

Composting can transform agricultural and municipal waste into a valuable soil amendment which increases SOC [7,8], while at the same time increasing soil nutrients and yields [9,10]. Both dairy manure [11] and food waste [12,13] have negative environmental effects, and composting offers one solution to recycle these wastes. For example, dairy manure compost applied at a rate of $105 \, \text{Mg DM ha}^{-1}$ increased SOC by 73% and supported corn yields similar to that of inorganic fertilizer [14]. In wheat, both municipal organic waste compost [9] and dairy manure compost [15] increased soil nutrient pools and yield. Recycled waste composts can increase SOC in almond production [16] and are applied by growers with the goals of increasing tree nutrition and beneficial soil biology [17].

Applying composts can increase microbial populations [18,19] and microbial diversity [20], which has sometimes been associated with increased nutrient use efficiency [21]. However, other studies have found slightly negative [22] or neutral [23] effects of compost on microbial diversity and activity. Terms such as high and low in this case are relative, though, since the minimum amount of biodiversity necessary to maintain plant health is often unknown [24]. Often it is not the raw number of species that is important, but rather the functions certain species perform [25], which in soil, includes organic matter decomposition and cycling nutrients [26]. Since many microbes do not grow well in the laboratory, their identity is only known through DNA sequencing, and directly linking natural populations to function requires a combination of genomic and culture-based approaches [27]. Although recent technological advances (such as lower costs of high throughput molecular sequencing) show promise, scientific understanding of how microbial diversity influences ecosystem functioning in agro ecosystems is still in its early stages [25,28].

Differences in microbial communities are reflected in bacterial and fungal-feeding nematodes, which respond rapidly to the abundance of their prey [29] and channel resources derived from bacterial and fungal decomposition [30]. For example, fungal-feeding nematodes proliferate with more processed resources [29,31,32], while increases in bacterial-feeders have been found with more readily decomposable resources such as compost feedstocks [31] and cover crops [33]. Nematodes have been proposed as particularly good indicators of soil health [34,35] because of their ubiquitous presence in soils, diverse number of functions that they provide, and their rapid response to changes in management. Previous studies have found that applying composted agricultural waste increased overall nematode biomass, as well as the abundance of bacterial-feeders [36], fungal-feeders and omnivores/predators [37]. However, one study has [38] observed no effect of composted waste on nematode communities.

To optimize outcomes for plant and soil health, greater understanding is needed about how management practices affect interrelationships between SOC, nutrient cycling and soil food webs [39]. The current study examined the effects of applying two recycled waste composts (incorporating either dairy manure or municipal food waste) in an almond orchard, comparing them to either a fertilized (N+) or unfertilized (N−) control over two years. The objectives were: (1) To determine the effects of composts on SOC and nitrogen (N) pools, (2) To characterize how composts influenced bacterial and nematode communities and their interrelationships, and (3) To determine if composts resulted in differences in plant growth and leaf nutrient content. We hypothesized that both composts would increase SOC and soil nutrient pools, with cascading effects on food webs and plant productivity. Expanding knowledge about the biological regulators of organic matter and nutrient dynamics could facilitate future management of food webs for increased soil fertility, which is particularly important in organic farming systems [39].

2. Materials and Methods

2.1. Orchard Establishment and Experimental Design

To characterize how soil communities responded to compost addition, an almond orchard was planted in March 2016 at the Armstrong Plant Pathology Research Station, University of California Davis, Davis, CA, USA. The soil was mapped as a Yolo silty clay loam [40] and contained 0.97% C and 0.1% N with a pH of 7.8. The experiment compared the effects of two commercially available composts. The first, termed food waste compost (FWC), incorporated municipal food scraps, yard clippings and agricultural waste. It had a C:N of 14 and was composed of 49.8% organic matter and 25% organic carbon with 1.8% N. The second compost included waste streams classified as agricultural, green waste and dairy manure, and will be referred to as dairy manure compost (DMC). This compost had a C:N of 10.8, and was composed of 28.5% organic matter and 14% organic carbon with 1.3% N.

The experiment had four main treatments: FWC, DMC, nitrogen fertilizer (N+) and a control without any organic or inorganic amendments (N−). All treatments were planted with container nursery stock of 'Nonpareil' almonds on 'Krymsky 86' rootstock on a 2.7 * 4.9 m spacing. Each experimental unit consisted of two trees separated from other treatments by one pollinizer buffer tree (either the almond cultivar, 'Monterey', on 'Krymsky 86' rootstock or 'Wood colony' on 'Krymsky 86'). There were six replicates of each treatment applied in a randomized complete block design, treating tree row as the block, so that within each of the six tree rows, each treatment was replicated once. Almond plantings were watered for approximately 20 h each week by drip irrigation, with each tree having two 7.6 L h^{-1} emitters.

Both composts were applied pre-planting with a front loader and spread evenly with shovels. Composts were applied at a rate of 112.09 metric tons dry weight ha^{-1} to an 8 m^2 area comprising the berms of two tree rows. This rate was chosen to approximate the estimated N needed by the trees in their first year; assuming that only 10% of the total N from the compost mineralized [41], so that each tree received a total of 0.08 kg N. For the N fertilizer treatment, urea ammonium nitrate (UAN-32) was applied in year one at a rate of 88.7 mL N per tree or 91.9 kg N ha^{-1}, spread out into six applications of 14.8 mL ($\frac{1}{2}$ oz) tree^{-1}. Applications occurred three times in June, twice in July and once in October of 2016. In 2017, the total applied N in fertilizer treatments increased to 177.4 mL N tree^{-1}, as recommended [42] with applications occurring three times in May and three times in June. Fertilizer was applied by injection into the irrigation lines, and separate lines were used for fertilizer, compost and control trees so that all treatments received equal amounts of water.

2.2. Soil Sampling and Plant Measurements

Soil was sampled with two 6.3 cm diameter cores at a depth of 0–25.4 cm, 30 cm from the trunk of each tree and composited for each plot replicate. Sampling occurred three times each year in May, July and October. Fresh soil samples were analyzed for mineral N contents using 2 M KCl extraction of 40 g soil followed by colorimetric determination of nitrate (NO_3^-) and ammonium (NH_4^+) contents [43]. After soil was dried at 60 °C, and sieved to 2 mm, soil particle sizes were determined by laser diffraction on a Beckman-Coulter LS-230 Particle Size Analyzer [44]. Finely ground soil was analyzed for total N (%) and C (%) on a Europe Hydra 20/20 isotope ratio mass spectrometer at the University of California Davis Stable Isotope Facility. Labile soil carbon, represented as permanganate oxidizable carbon (POXC) was measured in October of each year on finely ground soil following Culman et al. [45]. Briefly, triplicate samples of 2.5 g soil were oxidised with 0.02 mol L^{-1} KMnO4 with 2 min shaking followed by 10 min incubation and non-reduced Mn^{7+} quantified by colorimetry.

Indicators of tree productivity included trunk diameter and leaf N. Trunk diameters were measured with a caliper at the beginning and end of each year, in May and October, two feet above the soil. Leaves were collected for N analysis three times each year in May, July and October. For each of the two trees in each plot, five young, fully mature leaves were collected so that 10 leaves were collected for each experimental replicate. Leaves were dried at 60 °C for one week, finely ground, and total N (%) and C (%) determined on a Europe Hydra 20/20 isotope ratio mass spectrometer at the University of California Davis Stable Isotope Facility.

2.3. Nematode Communities

Nematodes were extracted from 200 mL of field moist soil using a sieving and decanting technique followed by sugar centrifugation [46]. The total number of nematodes in each sample was counted and the first 200 encountered on a slide were identified. Most nematodes were identified to the genus level [47], although some were only identified to the family level, such as those in the families Qudsianematidae and Tylenchidae, as genera within these groups are difficult to distinguish. The abundance of nematode groups identified were used to calculate indices of ecosystem functioning. For example, the Enrichment Index (EI) indicates the activity of primary detrital consumers. [48], while the Channel Index provides information on whether decomposition is proceeding more through bacterial or fungal channels, and the Structure index increases with food web complexity [48]. Nematode metabolic footprints were also calculated to provide an estimate of the contribution of different functional guilds of nematodes to functions related to carbon and nutrient cycling based on their size-dependent metabolic activity [30]. Calculations of indices and metabolic footprints were completed using the online platform, NINJA: 'Nematode INdicator Joint Analysis' [49].

2.4. Phylogenetic and Taxonomic Analysis of Prokaryotic Communities

Soils for molecular analysis (which were only collected in July and October of each year) were transported to the laboratory on ice and immediately stored at -80 °C until DNA extraction. Total DNA was extracted from 0.25 g of soil per sample using the DNeasy PowerLyzer PowerSoil kit (Qiagen, Inc., Germantown, MD, USA) following the manufacturer's protocol. Gel electrophoresis was used to assess quality of DNA after each extraction. Yields were assessed with a Qubit 3 fluorometer (ThermoFisher, Waltham, MA, USA) and extractions producing >15.0 ng μL^{-1} DNA were used to construct 16S rRNA gene libraries.

Libraries were prepared using a standard 16S rRNA primer pair: 515-F (GTGCCAG-CMGCCGCGGTAA) and 806-R (GGACTACHVGGGTWTCTAAT) targeting the gene's V4 hypervariable region ([50]). PCR was performed in duplicate using Phusion Hot Start II High-Fidelity PCR Master Mix (Thermo Scientific Inc., Waltham, MA). Reactions were conducted using a modified form of the manufacturer's protocol, with 1 μL DNA template (15 ng μL^{-1}), 1 μL of each primer (10 μMol), 10 μL master mix, and 7 μL water to reach a final volume of 20 μL reaction^{-1}. Negative controls were used in each batch of PCRs, substituting 1 μL DNA template with 1 μL water and a unique reverse barcode to remove contaminating DNA following sequencing analysis. All reactions were conducted using the C1000 Touch Thermo Cycler from Bio-Rad Laboratories, Inc. (Hercules, CA, USA). PCR cycles included a 30 s initial denaturation at 98 °C, followed by 27 cycles of denaturation at 98 °C for 10 s, annealing at 50 °C for 30 s, extension at 72 °C for 15 s, and a 7 min final extension at 72 °C before being held at 4 °C. Following PCR, a 3 μL aliquot of each reaction was assessed on an agarose gel to ensure specific and successful amplification. Duplicate reactions were then mixed and assessed for concentration using the Qubit 3 fluorometer (ThermoFisher, Waltham, MA, USA). Next, 100 ng of each successful reaction was pooled and purified using the QIAGEN's QIAquick PCR Purification Kit (Qiagen, Inc., Germantown, MD, USA) according to the manufacturer's protocol. Completed libraries were then sequenced on the MiSeq PE250 system at the UC Davis DNA Technologies Core and processed using the Dada2 platform using conventional methods recently described [51].

Diversity was quantified using both taxonomic- and phylogenetic-based methods. Taxonomic alpha diversity was measured as exact sequence variants (ESV) [52] and taxonomic group richness and equitability (Shannon diversity) within individual communities. Taxonomic dissimilarity of different communities was measured as the Bray-Curtis distance among samples based on ESV and taxonomic group membership [53]. Bacterial soil functions were inferred from taxonomy using FAPROTAX [54] which uses established literature on cultured strains to synthesize a putative functional profile for the total community.

2.5. Statistics

The statistical program R v4.0.3 (R Core Team, 2021) was used to assess the effects of compost on soil, bacteria, nematode and plant variables. Treatment effects were analyzed using analysis of variance (ANOVA), with means separated by Tukey's honestly significant difference (HSD) tests. Assumptions of homogeneity of variance and normality were assessed by Levene's and Shapiro–Wilk tests, respectively, and data were either log or square root transformed as needed. In cases where assumptions could not be met even with transformation, differences between treatments were assessed by non-parametric Kruskal-Wallace tests followed by a post-hoc Dunn's test. To measure changes in trunk diameter, which was taken for both trees in a plot, mixed effects analysis was performed using the R package lme4 [55] with treatment as a fixed effect and plot as a random effect. Relationships between trunk diameter and soil properties were examined for each timepoint using Pearson's correlations. Since nematode and bacterial abundance was often non-normally distributed, Spearman's rank correlations were used to determine the relationships between genera within each treatment.

For the Domains Bacteria and Archaea, the above analyses focused on the 20 most abundant taxa in the dataset. This assessment was verified using rank abundance curves where genera abundance was greatly diminished beyond the most represented taxa. All 797 genera were included, though, for non-metric multidimensional scaling (NMDS) analyses, which compared community composition between treatments. NMDS was conducted using the metaMDS function in the vegan package of R [56] and plotted using ggplot2 [57]. The vegan package was also used to calculate diversity indices (Shannon diversity, evenness and richness).

3. Results

3.1. Soil Variables

Compost treatments had higher total mineral soil N than controls throughout most of the experiment, but only influenced labile N pools within the first year (Table 1). Immediately after compost application in May of year one, FWC treated plots had higher NH_4^+-N than controls ($p < 0.01$) and a similar trend was observed for DMC ($p = 0.06$). Soon after it was applied through the irrigation system, fertilizer treatments had more than 3 times higher NO_3^--N and NH_4^+-N than controls in July (Table 1, $p = 0.01$). By October, though, the effects of the fertilizer had dissipated and DMC had increased NH_4^+-N and NO_3^--N to more than two times higher than controls or fertilizer treatments ($p < 0.05$). In the second year, the only effects seen in labile N pools were that fertilizer dramatically increased NO_3^--N compared to all other treatments soon after application in July (Table 1).

Composts application increased SOC measured at multiple timepoints (Table 1), often increasing total soil C by at least 50% ($p < 0.05$). Similar trends were seen in the more labile C pool, POXC. In year one, POXC levels were more than twice as high for FWC and DMC compared to controls ($p < 0.01$). Both composts also showed elevated POXC levels compared to fertilizer treatments ($p < 0.01$). By the end of the experiment, in October of year two, FWC and DMC continued to have higher POXC than either N− controls or fertilizer treatments ($p < 0.01$). Particle size analysis showed that soil had an average of 37.6% sand, 58.1% silt and 4.33% clay. In contrast to the trend for POXC, the percent clay content for FWC (3.77 ± 0.29) and DMC (3.60 ± 0.20) was lower than either controls (4.89 ± 0.18) or fertilizer treatments ($p < 0.05$; 5.06 ± 0.33).

Table 1. Average soil properties ± standard error of measurement (SEM) from an almond orchard receiving different organic amendments. Letters denote statistical differences of $p < 0.05$ determined by either post hoc Tukey's honestly significant difference (HSD) test or Dunn's test, for non-normally distributed data. DMC = Dairy manure compost, FWC = Food waste compost, N+ = nitrogen fertilizer, N− = unamended control. POXC = permanganate oxidizable carbon.

	NH_4^+ N (µg/g)				NO_3^- N (µg/g)				%N				%C				POXC (mg/kg)			
May 2016																				
DMC	6.49	±	1.93	abd	40.52	±	3.85		0.12	±	0.01		1.25	±	0.2					
FWC	10.45	±	1.51	d	39	±	4.41		0.14	±	0.03		1.57	±	0.4					
N+	1.31	±	0.36	ac	37.19	±	6.07		0.16	±	0.02		1.69	±	0.27					
N−	1.79	±	0.63	a	37.22	±	10.5		0.11	±	0.01		1.1	±	0.12					
July 2016																				
DMC	2.86	±	0.39	ab	15.28	±	3.15	ab	0.17	±	0.02	a	1.64	±	0.17	a				
FWC	2.97	±	0.53	ab	4.87	±	2.38	a	0.18	±	0.03	a	2.08	±	0.45	a				
N+	8.04	±	2.5	b	22.19	±	1.34	b	0.1	±	0	b	0.94	±	0.02	b				
N−	2.09	±	0.25	a	6.45	±	2.45	a	0.1	±	0.01	b	0.99	±	0.05	b				
October 2016																				
DMC	2.69	±	0.39	b	39.38	±	11.87	a	0.19	±	0.02	a	1.7	±	0.17	a	700.63	±	43.96	b
FWC	1.52	±	0.64	ac	24.93	±	6.34	ac	0.21	±	0.03	a	2.17	±	0.33	a	855.34	±	104.51	b
N+	0.58	±	0.06	a	5.28	±	0.73	b	0.11	±	0.01	b	1.05	±	0.1	b	263.45	±	33.42	a
N−	0.73	±	0.23	a	12.55	±	5.59	bc	0.1	±	0	b	0.97	±	0.03	b	291.77	±	33.15	a
May 2017																				
DMC	0.48	±	0.23		3.98	±	1.06		0.17	±	0.03	a	1.56	±	0.25	ac				
FWC	0.75	±	0.26		3.55	±	1.48		0.35	±	0.07	a	3.9	±	0.8	a				
N+	0.26	±	0.12		1.88	±	0.54		0.1	±	0	b	0.96	±	0.03	bc				
N−	0.2	±	0.1		2.68	±	0.41		0.1	±	0	b	0.9	±	0.02	bc				
July 2017																				
DMC	1.18	±	0.54		3.38	±	1.78	a	0.18	±	0.02	a	1.72	±	0.22	a				
FWC	0.46	±	0.4		4.71	±	2.76	a	0.17	±	0.03	a	1.82	±	0.33	a				
N+	12.23	±	7.94		43.4	±	12.77	b	0.11	±	0	b	0.96	±	0.04	b				
N−	0.26	±	0.17		3	±	3.44	a	0.1	±	0	b	0.93	±	0.03	b				
October 2017																				
DMC	0.76	±	0.32		4.74	±	2.35		0.15	±	0	a	1.32	±	0.05	a	551.01	±	43.71	b
FWC	0.49	±	0.14		3	±	1.25		0.14	±	0.01	a	1.35	±	0.11	a	600.92	±	61.12	b
N+	1.03	±	0.7		2.92	±	0.97		0.11	±	0.01	b	0.99	±	0.08	b	377.45	±	17.88	a
N−	0.27	±	0.24		2.81	±	0.64		0.1	±	0	b	0.9	±	0.02	b	281.56	±	45.21	a

3.2. Bacterial and Archaeal Communities

Sequencing identified 797 bacterial and archaeal genera, whose community composition differed between treatments over time (Figure 1). Across all treatments, bacterial species richness (determined by exact sequence variants) and Shannon diversity generally decreased in year two compared to year one (Table 2, $p < 0.01$). While treatments did not influence species richness, both composts increased species evenness compared to fertilizer and control treatments ($p < 0.05$, Table 2) at certain time points. In July of year two, fertilizer treatments decreased Shannon diversity compared to DMC and FWC ($p < 0.05$). Non-metric multidimensional scaling analysis showed that in July and October of year one, FWC and DMC treatments hosted communities that were distinct, both from each other, and from the control and fertilizer treatments (Figure 1). These differences had largely disappeared by year two, but DMC again clustered slightly apart from other treatments by the end of the experiment. Differential abundance analysis showed that several taxa responded negatively to compost application (Tables S1 and S2) including *Rubrobacter*, *Pseudarthrobacter* and *Solirubacter*. Adding FWC increased the abundance of *Lysinibacillus* compared to either controls ($p < 0.01$) or nitrogen treatments ($p < 0.01$) in July of year one. Both composts increased *Steroidobacter* compared to nitrogen treatments in October of year two ($p < 0.01$).

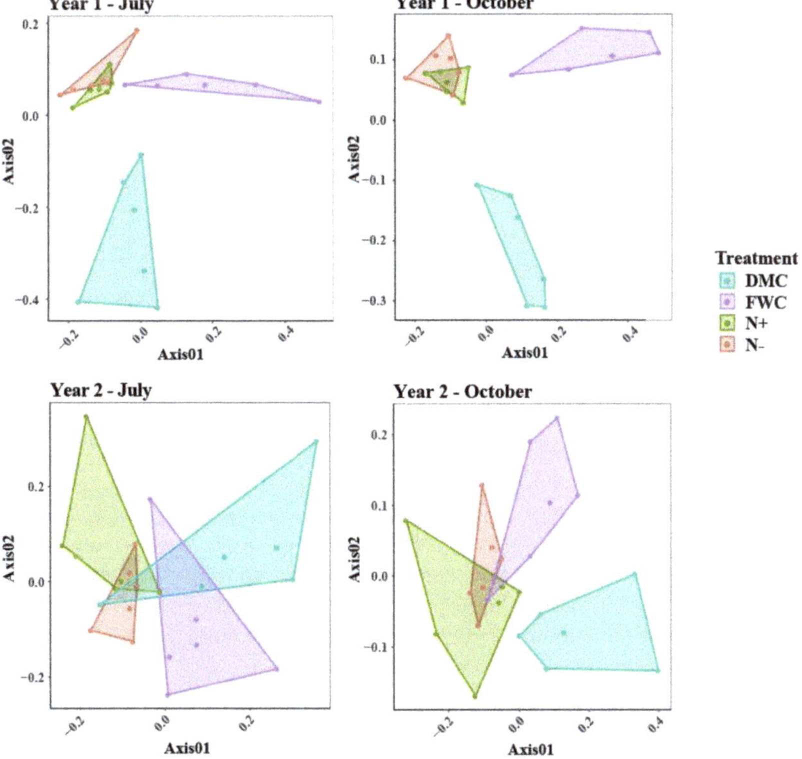

Figure 1. Non-metric multidimensional scaling analysis (NMDS) of 797 bacterial and archaeal genera isolated from almond orchard soil in two years following compost amendment application. DMC = Dairy manure compost, FWC= Food waste compost, N+ = nitrogen fertilizer, N− unamended control.

Table 2. Average bacterial and archaeal species richness, evenness and Shannon Diversity indices ± standard error of measurement (SEM) from an almond orchard receiving different organic amendments. Letters denote statistical differences of $p < 0.05$ determined by either post hoc Tukey's honestly significant difference (HSD) test or Dunn's test, for non-normally distributed data. DMC = Dairy manure compost, FWC = Food waste compost, N+ = nitrogen fertilizer, N− = unamended control. Categories with 0 values indicate SEM under 0.01.

	Richness			Evenness			Shannon Diversity		
July 2016									
DMC	1071.00	±	64.64	0.87	±	0.01	6.07	±	0.08
FWC	1067.33	±	25.81	0.88	±	0.00	6.14	±	0.03
N+	998.00	±	82.25	0.87	±	0.01	5.96	±	0.11
N−	1011.00	±	68.66	0.87	±	0.00	5.99	±	0.05
October 2016									
DMC	944.67	±	38.31	0.88	±	0.00 a	6.06	±	0.05
FWC	881.33	±	61.41	0.89	±	0.00 a	5.99	±	0.07
N+	1009.33	±	25.92	0.87	±	0.00 b	6.01	±	0.03
N−	1044.00	±	52.56	0.87	±	0.00 b	6.02	±	0.06
July 2017									
DMC	781.33	±	65.28	0.89	±	0.01 a	5.94	±	0.08 a
FWC	789.67	±	55.18	0.90	±	0.00 a	5.97	±	0.05 a
N+	659.67	±	55.71	0.87	±	0.00 b	5.65	±	0.08 b
N−	779.00	±	45.29	0.87	±	0.00 b	5.79	±	0.05 ab
October 2017									
DMC	798.17	±	38.12	0.89	±	0.00 a	5.94	±	0.04
FWC	716.00	±	68.84	0.88	±	0.00 ab	5.77	±	0.10
N+	739.00	±	66.23	0.87	±	0.00 b	5.71	±	0.09
N−	772.50	±	53.37	0.86	±	0.01 b	5.72	±	0.08

When bacteria and archaea were separated into groups indicative of function, compost application showed a higher relative abundance of those with carbon degrading potential (Figure 2). Both organic treatments increased the relative abundance of bacteria with xylanolytic potential compared to control and fertilizer treatments in July and October of year one ($p < 0.01$). For bacteria with cellulolytic potential, only DMC caused increases, which were three times higher than N− controls in July of year one ($p < 0.01$) and 49 times higher than controls in October ($p < 0.01$). Effects were less pronounced in the second year, although FWC continued to have slightly higher xylanolytic potential than controls in July ($p < 0.01$) and October ($p = 0.05$) and DMC had higher cellulolytic potential than controls ($p < 0.01$). Some differences were also observed between the two sources of compost. FWC had higher abundance of bacteria with xylanolytic potential than DMC in October of year one ($p = 0.02$), but DMC had higher abundances of bacteria with cellulolytic potential ($p < 0.01$), a trend which continued into both timepoints of the second year ($p = 0.02$; $p < 0.01$).

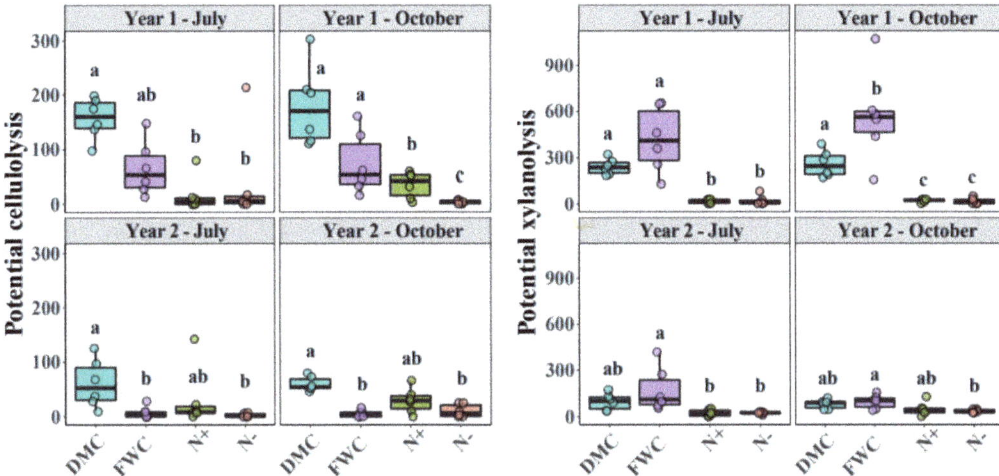

Figure 2. The relative abundance of bacteria with presumptive cellulolytic and xylanolytic potential from almond orchard soil in two years following compost amendment application. N = nitrogen fertilizer, C = control, DMC = Dairy manure compost, FWC = food waste compost. Letters denote statistical differences of $p < 0.05$ determined by either post hoc Tukey's honestly significant difference (HSD) test.

3.3. Nematode Communities

Over the two years of the experiment, 15 groups of nematodes were identified (Tables 3 and 4) including bacterial feeders, fungal feeders, plant root feeders, omnivores, and predators. Nematodes were very abundant, with an average of 1975.6 ± 108.5 individuals 200 mL^{-1} soil and 1.3 ± 0.1 mg estimated biomass. While some nematodes, such as *Panagrolaimus* and *Aphelenchoides*, were common throughout the experiment, others, such as the bacterial feeder, *Prismatolaimus*, were not detected in any samples until the fall of year one. After the initial disturbance of planting, the complexity of the nematode food web increased over time, with higher levels of the Structure Index ($p < 0.01$, F = 13.6), and Structure metabolic footprint ($p = 0.02$, F = 6.3) in the second year (*data not shown*).

Different groups of bacterial feeding nematodes responded to compost treatments over time (Table 3). FWC treated plots had greater relative abundances of *Cephalobus* compared to control or N plots ($p < 0.01$) after composts were incorporated in May of year one. DMC similarly increased *Cephalobus* compared to N treatments ($p = 0.02$), although these effects were observed before fertilizer treatments had been applied (Table 3). By the end of year one in October, *Prismatolaimus* made up a larger portion of the nematode community in DMC compared to N treatments ($p = 0.03$) although the abundance of these nematodes was generally low (under 6%). In May of year two, the total relative abundance of bacterial feeding nematodes was greater in DMC treatments (Table 4) compared to N+ treatments ($p < 0.01$).

In contrast to the effect seen for bacterial feeders, composts decreased the relative abundance of certain fungal feeding nematodes (Table 4). In spring of year two, both DMC and FWC treatments depressed the relative abundance of *Aphelenchus* compared to control and N treatments ($p < 0.05$). This contributed to lower abundance of fungal feeders, overall, in compost treatments compared to N or control treatments ($p < 0.04$). DMC continued to depress *Aphelenchus* abundance into the summer of year two compared to N treatments ($p = 0.04$, Table 5). However, no effects were seen on other nematode indicators such as the Channel index or nematode Fungal metabolic footprint (*data not shown*).

Table 3. The average relative abundance of individual nematode groups in year one from an almond orchard receiving different organic amendments. Letters denote statistical differences of $p < 0.05$ determined by Tukey's honestly significant difference (HSD) test. DMC = Dairy manure compost, FWC = Food waste compost, N+ = nitrogen fertilizer, and N− = unamended control. For trophic groupings of nematodes bact. = bacterial feeders, fung. = fungal feeders, omn. = omnivores, pred. = predators, and herb. = root herbivores. Categories with 0 values indicate relative abundances under 0.01 (1%).

		May 2016			
Nematode taxa		**DMC**	**FWC**	**N+**	**N−**
Panagrolaimus	bact.	0.40	0.48	0.36	0.39
Mesorhabditis	bact.	0.12	0.09	0.05	0.08
Cephalobus	bact.	0.06 ab	0.10 b	0.01 c	0.02 a
Acrobeloides	bact.	0.07	0.00	0.08	0.07
Acrobeles	bact.	0.00	0.00	0.00	0.00
Prismatolaimus	bact.	0.01	0.00	0.02	0.00
Aphelenchoides	fung.	0.12	0.15	0.16	0.17
Aphelenchus	fung.	0.09	0.06	0.12	0.09
Discolaimus	pred.	0.00	0.00	0.00	0.00
Qudsianematidae	omn.	0.01	0.00	0.04	0.02
Mesodorylaimus	omn.	0.00	0.00	0.00	0.00
Tylenchidae	herb.	0.11	0.09	0.14	0.15
Pratylenchus	herb.	0.02	0.01	0.02	0.00
Total bacterial feeders		0.65	0.68	0.52	0.57
Total fungal feeders		0.21	0.22	0.28	0.26
Total herbivores		0.13	0.10	0.16	0.15
Total omnivores		0.01	0.00	0.04	0.02
		July 2016			
Nematode taxa		**DMC**	**FWC**	**N+**	**N−**
Panagrolaimus	bact.	0.09	0.14	0.12	0.15
Mesorhabditis	bact.	0.22	0.19	0.18	0.23
Acrobeloides	bact.	0.10	0.10	0.11	0.10
Acrobeles	bact.	0.01	0.01	0.00	0.00
Prismatolaimus	bact.	0.00	0.00	0.00	0.00
Aphelenchoides	fung.	0.22	0.19	0.18	0.18
Aphelenchus	fung.	0.12	0.11	0.10	0.14
Discolaimus	pred.	0.00	0.00	0.00	0.00
Qudsianematidae	omn.	0.01	0.01	0.02	0.02
Mesodorylaimus	omn.	0.01	0.00	0.00	0.00
Tylenchidae	herb.	0.19	0.22	0.22	0.17
Pratylenchus	herb.	0.03 ab	0.03 ab	0.07 b	0.00 a
Total bacterial feeders		0.48	0.43	0.40	0.43
Total fungal feeders		0.28	0.30	0.29	0.37
Total herbivores		0.22	0.25	0.29	0.17
Total omnivores		0.02	0.01	0.02	0.02
		October 2016			
Nematode tax		**DMC**	**FWC**	**N+**	**N−**
Panagrolaimus	bact.	0.10	0.11	0.14	0.11
Mesorhabditis	bact.	0.11	0.09	0.11	0.09
Acrobeloides	bact.	0.07	0.09	0.08	0.08
Acrobeles	bact.	0.09	0.02	0.00	0.00
Prismatolaimus	bact.	0.05 a	0.04 ab	0.01 b	0.04 ab
Aphelenchoides	fung.	0.10	0.12	0.10	0.13
Aphelenchus	fung.	0.07	0.07	0.10	0.07
Discolaimus	pred.	0.00	0.00	0.01	0.00
Qudsianematidae	omn.	0.00	0.00	0.00	0.00
Mesodorylaimus	omn.	0.03	0.02	0.03	0.04
Tylenchidae	herb.	0.37	0.44	0.40	0.42
Pratylenchus	herb.	0.01	0.01	0.03	0.02
Total bacterial feeders		0.41	0.34	0.34	0.32
Total fungal feeders		0.17	0.19	0.20	0.19
Total herbivores		0.38	0.45	0.43	0.44
Total omnivores		0.03	0.02	0.03	0.04

Table 4. The average relative abundance of individual nematode groups in year two from an almond orchard receiving different organic amendments. Letters denote statistical differences of $p < 0.05$ determined by post hoc Tukey's honestly significant difference (HSD) test. DMC = Dairy manure compost, FWC = Food waste compost, N+ = nitrogen fertilizer, and N− = unamended control. For trophic groupings of nematodes bact. = bacterial feeders, fung. = fungal feeders, pred. = predators, omn. = omnivores, and herb. = root herbivores. Categories with 0 values indicate relative abundances under 0.01 (1%).

		May 2017			
Nematode taxa		**DMC**	**FWC**	**N+**	**N−**
Panagrolaimus	bact.	0.01	0.00	0.01	0.01
Mesorhabditis	bact.	0.01	0.02	0.02	0.01
Cephalobus	bact.	0.00	0.00	0.00	0.01
Eucephalobus	bact.	0.00	0.00	0.00	0
Acrobeloides	bact.	0.22	0.16	0.11	0.2
Acrobeles	bact.	0.00	0.03	0.00	0
Prismatolaimus	bact.	0.23	0.18	0.14	0.16
Aphelenchoides	fung.	0.12	0.12	0.09	0.11
Aphelenchus	fung.	0.07 b	0.07 b	0.20 a	0.2 a
Discolaimus	pred.	0.00	0.00	0.01	0.00
Qudsianematidae	omn.	0.03	0.05	0.11	0.08
Dorylaimus	omn.	0.00	0.01	0.01	0.01
Tylenchidae	herb.	0.29	0.33	0.28	0.2
Meloidogyne	herb.	0.01	0.03	0.01	0.00
Total bacterial feeders		0.48 a	0.38 ab	0.28 b	0.39 ab
Total fungal feeders		0.19 b	0.19 b	0.30 a	0.31 a
Total herbivores		0.16 ab	0.19 b	0.15 ab	0.1 a
Total omnivores		0.03	0.06	0.12	0.09
		July 2017			
Nematode taxa		**DMC**	**FWC**	**N+**	**N−**
Panagrolaimus	bact.	0.02	0.01	0.06	0.06
Mesorhabditis	bact.	0.01	0.01	0.01	0.01
Eucephalobus	bact.	0.01	0.00	0.02	0.01
Acrobeloides	bact.	0.25	0.13	0.19	0.17
Prismatolaimus	bact.	0.09	0.13	0.03	0.10
Aphelenchoides	bact.	0.01	0.01	0.03	0.03
Aphelenchus	fung.	0.13 a	0.16 ab	0.30 b	0.29 ab
Qudsianematidae	omn.	0.01	0.02	0.01	0.02
Dorylaimus	omn.	0.01	0.02	0.01	0.03
Tylenchidae	herb.	0.46 ab	0.49 b	0.33 ab	0.27 a
Total bacterial feeders		0.38	0.29	0.31	0.34
Total fungal feeders		0.13 b	0.18 ab	0.33 c	0.32 a
Total herbivores		0.23 ab	0.25 b	0.17 ab	0.14 a
Total omnivores		0.02	0.04	0.03	0.05
		October 2017			
Nematode taxa		**DMC**	**FWC**	**N+**	**N−**
Panagrolaimus	bact.	0.14	0.15	0.09	0.13
Mesorhabditis	bact.	0.09	0.10	0.11	0.06
Cephalobus	bact.	0.17	0.14	0.18	0.15
Prismatolaimus	bact.	0.08	0.11	0.06	0.15
Aphelenchoides	fung.	0.16	0.15	0.15	0.22
Aphelenchus	fung.	0.00	0.01	0.01	0.01
Microdorylaimus	omn.	0.01	0.01	0.02	0.00
Tylenchidae	herb.	0.34	0.33	0.31	0.25
Paratylenchus	herb.	0.00	0.00	0.00	0.01
Tylenchorhynchus	herb.	0.00	0.00	0.01	0.01
Pratylenchus	herb.	0.00	0.01	0.05	0.01
Total bacterial feeders		0.48	0.49	0.44	0.49
Total fungal feeders		0.16	0.16	0.16	0.23
Total herbivores		0.17	0.18	0.22	0.15
Total omnivores		0.01	0.01	0.02	0.00

Table 5. Average leaf nutrients, trunk diameter and growth ± standard error of measurement (SEM) from an almond orchard receiving different organic amendments. Letters denote statistical differences of $p < 0.05$ determined by either post hoc Tukey's honestly significant difference (HSD) test or Dunn's test, for non-normally distributed data. DMC = Dairy manure compost, FWC = Food waste compost, N+ = nitrogen fertilizer, N− = unamended control.

	% N			% C			Trunk Diameter (mm)			Diameter Increase (mm)		
May 2016												
DMC	3.14	±	0.09	44.16	±	0.20	9.40	±	0.46			
FWC	3.11	±	0.15	43.85	±	0.61	9.70	±	0.22			
N+	2.97	±	0.17	44.29	±	0.13	10.22	±	0.23			
N−	3.09	±	0.03	44.02	±	0.14	10.16	±	0.26			
July 2016												
DMC	2.92	±	0.15	43.85	±	0.23						
FWC	3.08	±	0.21	43.80	±	0.17						
N+	3.07	±	0.13	43.97	±	0.18						
N−	2.82	±	0.13	43.62	±	0.17						
October 2016												
DMC	3.22	±	0.06 ab	45.22	±	0.47	23.69	±	0.41	14.30	±	0.57 ab
FWC	3.24	±	0.14 ab	44.98	±	0.38	25.34	±	1.19	15.64	±	1.15 a
N+	3.46	±	0.19 a	45.11	±	0.21	22.61	±	0.68	12.39	±	0.70 ab
N−	2.86	±	0.10 b	44.14	±	0.16	21.48	±	1.05	11.32	±	0.88 b
May 2017												
DMC	2.66	±	0.09	46.07	±	0.17						
FWC	2.51	±	0.08	45.54	±	0.27						
N+	2.47	±	0.05	45.23	±	0.48						
N−	2.58	±	0.08	46.00	±	0.36						
July 2017												
DMC	2.33	±	0.07 b	45.96	±	0.26						
FWC	2.19	±	0.05 b	45.76	±	0.42						
N+	2.84	±	0.06 a	45.41	±	0.27						
N−	2.11	±	0.08 b	45.28	±	0.27						
October 2017												
DMC							47.05	±	2.10	37.65	±	1.98
FWC							48.56	±	2.22	38.86	±	2.29
N+							48.27	±	3.48	38.05	±	3.53
N−							42.68	±	2.01	32.52	±	2.02

Herbivorous nematodes increased with both N and FWC treatments compared to controls, although these effects occurred in different timeframes (Tables 3 and 4). In July of year one, recently fertilized plots had a higher relative abundance of the plant parasitic nematode, *Pratylenchus*, than controls (Table 3; $p < 0.01$). However, compost treatments did not influence herbivorous nematodes until the following spring (Table 4), after which the relative abundance of herbivores was higher in FWC treatments compared to controls in both May ($p = 0.03$) and July ($p = 0.03$) of year two. This was particularly apparent for root tip feeding nematodes in the family *Tylenchidae*, which were more abundance with FWC than controls ($p = 0.03$).

3.4. Relationships between Microbes and Nematodes

When the relationship between bacterial and nematode groups were examined, some groups showed consistent trends across all treatments, while others showed relationships that were more treatment specific (Figure 3). For example, in all treatment categories, the bacterial genera *Bacillus* was positively associated with bacterial feeding nematodes such as *Panagrolaimus*, *Rhabditis* and *Cephalobus*, as well as the fungal feeder, *Aphelenchoides* ($p < 0.05$), but was negatively associated with the bacterial feeder, *Acrobeloides* ($p < 0.05$). In some cases, the addition of compost caused new relationships to become apparent (Figure 3). For example, *Lysinbacillus* showed a positive correlation with *Panagrolaimus* in FWC (rs = 0.41, $p < 0.05$) and DMC treatments (rs = 0.47, $p < 0.05$) but had no relationship to this nematode genus in N− control (rs = 0.10) or N+ treatments (rs = 0.11). Other positive correlations between microbes and nematodes unique to the compost treatments included *MND1* and *Acrobeloides*, as well as *MND1*, *Bryobacter* and *Psychroglaciecola* positively associating with *Primatolaimus*, which was also associated with the archaea *Candidatus Nitrososphaera* ($p < 0.05$).

3.5. Plant Measurements

Almond trees increased in trunk diameter each year (Table 5, $p < 0.01$). However, slight differences in growth between the treatments were only observed in the first year (Table 5), with FWC treatments increasing trunk diameter by an average of 4.3 mm more than N− controls ($p = 0.02$). Increased growth in year one was positively associated with soil factors such as soil N ($p = 0.02$, R = 0.45) and POXC ($p < 0.01$, R = 0.57). Fertilizer increased leaf N contents compared to controls in October of year one ($p = 0.01$, Table 5). In July of year two, Fertilizer again increased leaf N compared to controls as well as FWC and DMC treatments ($p < 0.01$). The leaf N content of DMC treatments in July was also slightly higher than N− controls ($p = 0.08$). Leaf samples for October 2017 molded and were therefore not able to be analyzed. By the end of the experiment, cumulative growth for all trees was similar between treatments.

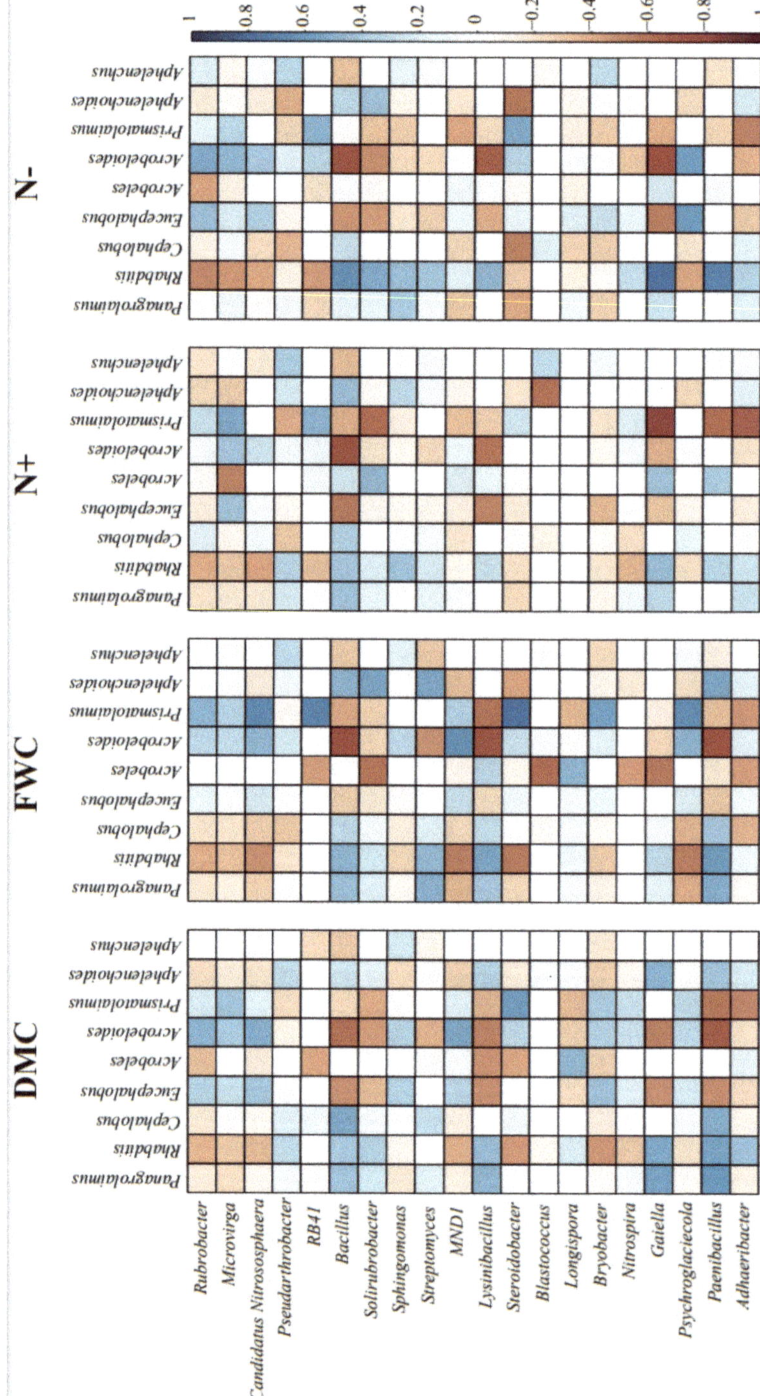

Figure 3. Heat map of Spearman rank correlation coefficients between nematode and bacterial and archaeal genera collected from almond orchard soil in 2016 and 2017 where treatments applied were either DMC = Dairy manure compost, FWC = Food waste compost, N+ = nitrogen fertilizer, or N− = unamended control. Only those relationships statistically significant at the $p < 0.05$ level are shown. Darker colors indicated stronger correlations, with blue indicating positive correlations and red negative.

4. Discussion

In this study, compost influenced microbial and nematode communities as well as soil organic matter, nutrient pools and plant growth. Compost released N in plant available forms within the first year, but in contrast to fertilizer, it did not increase leaf N concentrations, perhaps because the timing of release was asynchronous with plant needs [10]. Compared to N− controls, soil NO_3^- and NH_4^+ concentrations were only elevated with composts immediately after application and in the fall of year one, while fertilizer showed more consistent effects each year during the period it was applied in May–July. These results are partially in line with others who have found that composted waste products can increase orchard tree growth and that while slight increases in leaf nutrient content are possible, they are not as great as that seen with fertilizer [58–60]. The observed increase in tree growth with FWC compared to controls may have been due to increased root production, since higher populations of root feeding nematodes were also present in these treatments. Both composts altered soil properties, increasing SOC pools and reducing clay content, which could have made it physically easier for roots to penetrate the soil, however; effects on growth were only seen with FWC. Compared to DMC, FWC had a higher total N content, as well as higher organic matter content, which may have contributed to differences in tree growth.

In tandem with its plant and soil effects, compost influenced microbial communities within the first year of application. NMDS results showed that both composts temporarily shifted bacteria and archaea into separate, distinct communities from fertilized or unamended controls. In apples, composted poultry litter and yard waste has also been found to result in distinct bacterial communities [60]. Similar to Sharaf et al. [60], both composts in our study caused slightly higher Shannon diversity indices than fertilizer in year two, although this could be due to fertilizer suppressing microbes rather than compost elevating them [61]. While composts did not increase or decrease bacterial species richness compared to untreated controls, they did increase species evenness. Similar increases in bacterial evenness have been found with long term applications of manure and increased bacterial evenness has been shown to improve N cycling under stressful conditions, likely since many similar species are abundant enough to perform the same function [62].

Compost application especially increased those groups associated with carbon processing. Bacteria with presumptive cellulolytic and xylanolytic potential were more abundant with composts than fertilizer or untreated controls in the first year, suggesting that these groups were contributing to the observed community shifts. When the composts were compared to each other, FWC had higher abundance of bacteria with xylanolytic potential, while DMC had a higher abundance of bacteria with cellulolytic potential, perhaps due to differences in compost feedstocks. In another study [60], yard waste compost similarly increased bacterial groups implicated in generalized carbon cycling. Several of the specific taxa that increased with compost in this study are known to be associated with cycling nutrients. For example, *Steroidobacter* has been found to increase with organic amendment addition in soils with low initial SOC [63] and may be involved in nitrogen cycling under these conditions since it can only accept nitrates from a narrow range of compounds [64]. Species of *Lysinibacillus* have also been found to promote plant growth and enhance nutrient cycling [65–67]. Although they were not an explicit focus of this study, no known human pathogens were detected, which is a serious concerns for growers considering applying compost in almond orchards [17], however to confirm food safety, more targeted molecular approaches would likely be necessary.

Compost affected bacterial-feeding nematodes most strongly in the spring after application, with DMC increasing their relative abundance compared to untreated controls as well as fertilizer treatments. Since the rate of N mineralization from composts is likely to be slower than other organic amendments, with little N available in the first year, applications before the winter are sometimes recommended so that nutrients are available the following spring [68,69]. Although microbial communities were not measured at this time point, increased mineralization of nutrients from the compost may have stimulated bacterial growth, which could have provided food for the nematodes. Supporting this hypothesis is the observation that DMC plots had larger pools of NH_4^+-N the previous fall, although no differences in labile N were seen in spring of year two. Other studies have also observed increases in bacterial feeding nematodes with organic amendments, although effects vary with amendment composition [59,70]. It is surprising, though, that unlike microbial communities, compost did not induce large shifts in the species composition of nematodes. Herren et al. [38] also did not find changes in nematode community structure with compost addition, and suggested that recent tillage may disturbed the nematode community. Prior to planting with almonds, the field (which had been fallow for several years), was tilled, a practice known to decrease nematode community structure and alter the way soil food webs interact with organic amendments [71,72].

Relationships between nematode and bacterial/archaeal genera varied between treatments, suggesting that compost alters trophic dynamics between both groups. For example, the bacteria *Lysinbacillus*, had higher relative abundance with FWC than either N+ or N− controls. In correlations, *Lysinbacillus* was also associated with the bacterial feeding nematode, *Panagrolaimus*, but this relationship was only apparent within the compost treatments. It is known that bacteria can influence nematode survival and reproduction [73,74] and that nematodes can also alter microbial communities through their grazing [75,76], decreasing microbial biomass, but also increasing microbial activity [75]. Such predation can influence plant nutrient update [77,78], which may have contributed to the increased tree growth seen with the FWC treatment. Others have found that interactions between microbes and microbial feeding nematodes can vary with organic amendment application [70] as well as with their physical location in soil pores [79,80], so it is possible that by altering the composition of microaggregates in the soil, organic amendments influenced potential predator prey relationships. Since bacterial abundance was not directly quantified in this study, however; it is difficult to ascertain which of these mechanisms was the cause of the observed relationships.

5. Conclusions

In the current study, the two recycled waste composts increased SOC, but showed different effects on soil nitrogen pools and food webs. While DMC increased NH_4^+-N and NO_3^--N late in the first year and stimulated the activity of bacterial feeding nematodes, FWC was associated with more rapid increases in tree growth and populations of root feeding nematodes, as well as greater relative abundance of the bacteria, *Lysinbacillus*. Relationships between nematode and bacterial/archaeal genera varied between treatments, suggesting that compost can alter trophic interactions in the soil food web under field conditions, in contrast to previous studies that have usually used microcosms [75–77]. Compost applications influenced the soil food web, N cycling and tree growth mostly in the first year. In the second year, fertilizer showed greater effects than other treatments on tree growth and leaf N. Results suggest that while compost can contribute to biologically based nitrogen cycling and stimulate soil food webs, additional N inputs are likely needed to plant growth requirements.

Supplementary Materials: The following are available online at https://www.mdpi.com/article/10.3390/agronomy11091745/s1, Table S1: Differences in bacterial and archaeal (*Candidatus* Nitrososphaera) taxa between treatments compared to untreated controls based on ANOVA comparisons of relative abundance. Table S2: Differences in bacterial and archaeal (*Candidatus* Nitrososphaera) taxa between other treatments compared to the fertilizer treatment, based on ANOVA comparisons of relative abundance.

Author Contributions: Conceptualization, A.K.H. and J.M.S.; methodology, A.K.H., J.M.S. and M.C.C.P.L.; resources, J.L.M.R.; writing—original draft preparation, A.K.H.; writing—review and editing, J.M.S., J.L.M.R. and M.C.C.P.L. All authors have read and agreed to the published version of the manuscript.

Funding: This publication was supported by the U.S. Department of Agriculture's (USDA) Agricultural Marketing Service through Grant 17-SCBGP-CA-0041. Its contents are solely the responsibility of the authors and do not necessarily represent the official views of the USDA. This work was also supported in part by California Safe Soils (a producer of food hydrolysate).

Institutional Review Board Statement: Not applicable.

Informed Consent Statement: Not applicable.

Data Availability Statement: Data available upon request.

Acknowledgments: Special thanks to Bryan Pellissier for assistance with site management and agronomic advice, Xuyun Yang for assistance with nematode extraction and identification, and Janina Milkereit, Danica Dito, Andrew Cicchetto, Kisna Prado and Malina Loeher for field and lab assistance.

Conflicts of Interest: The authors declare no conflict of interest.

References

1. Stott, D.E. *Recommended Soil Health Indicators and Associated Laboratory Procedures*; Soil Health Technical Note No. 450-03; United States Department of Agriculture, Natural Resources Conservation Service: Washington, DC, USA, 2019.
2. Van Bruggen, A.; Semenov, A. In search of biological indicators for soil health and disease suppression. *Appl. Soil Ecol.* **2000**, *15*, 13–24. [CrossRef]
3. Pankhurst, C.; Doube, B.; Gupta, V. *Biological Indicators of Soil Health*; CAB Iternational: Wallingford, UK, 1997; Volume 17.
4. Lal, R. Soil health and carbon management. *Food Energy Secur.* **2016**, *5*, 212–222. [CrossRef]
5. Stevenson, F.J. *Humus Chemistry: Genesis, Composition, Reactions*; Wiley: New York, NY, USA, 1994.
6. Lal, R. Soil carbon sequestration to mitigate climate change. *Geoderma* **2004**, *123*, 1–22. [CrossRef]
7. Bernal, M.P.; Alburquerque, J.A.; Moral, R. Composting of animal manures and chemical criteria for compost maturity assessment. A review. *Bioresour. Technol.* **2009**, *100*, 5444–5453. [CrossRef] [PubMed]
8. Scotti, R.; Pane, C.; Spaccini, R.; Palese, A.M.; Piccolo, A.; Celano, G.; Zaccardelli, M. On-farm compost: A useful tool to improve soil quality under intensive farming systems. *Appl. Soil Ecol.* **2016**, *107*, 13–23. [CrossRef]
9. Duong, T.T.T.; Penfold, C.; Marschner, P. Amending soils of different texture with six compost types: Impact on soil nutrient availability, plant growth and nutrient uptake. *Plant Soil* **2012**, *354*, 197–209. [CrossRef]
10. Evanylo, G.; Sherony, C.; Spargo, J.; Starner, D.; Brosius, M.; Haering, K. Soil and water environmental effects of fertilizer-, manure-, and compost-based fertility practices in an organic vegetable cropping system. *Agric. Ecosyst. Environ.* **2008**, *127*, 50–58. [CrossRef]
11. Owen, J.J.; Silver, W.L. Greenhouse gas emissions from dairy manure management: A review of field-based studies. *Glob. Change Biol.* **2015**, *21*, 550–565. [CrossRef]
12. Levis, J.W.; Barlaz, M.A.; Themelis, N.J.; Ulloa, P. Assessment of the state of food waste treatment in the United States and Canada. *Waste Manag.* **2010**, *30*, 1486–1494. [CrossRef]
13. Adhikari, B.; Barrington, S.; Martinez, J. Urban food waste generation: Challenges and opportunities. *Int. J. Environ. Waste Manag. Indersci.* **2011**, *3*, 4–21. [CrossRef]
14. Butler, T.J.; Han, K.J.; Muir, J.P.; Weindorf, D.C.; Lastly, L. Dairy manure compost effects on corn silage production and soil properties. *Agron. J.* **2008**, *100*, 1541–1545. [CrossRef]
15. Butler, T.J.; Muir, J.P. Dairy manure compost improves soil and increases tall wheatgrass yield. *Agron. J.* **2006**, *98*, 1090–1096. [CrossRef]
16. Lepsch, H.C.; Brown, P.H.; Peterson, C.A.; Gaudin, A.C.M.; Khalsa, S.D.S. Impact of organic matter amendments on soil and tree water status in a California orchard. *Agric. Water Manag.* **2019**, *222*, 204–212. [CrossRef]
17. Khalsa, S.D.S.; Brown, P.H. Grower analysis of organic matter amendments in California orchards. *J. Environ. Qual.* **2017**, *46*, 649–658. [CrossRef]
18. Lee, J.J.; Park, R.D.; Kim, Y.W.; Shim, J.H.; Chae, D.H.; Rim, Y.S.; Sohn, B.K.; Kim, T.H.; Kim, K.Y. Effect of food waste compost on microbial population, soil enzyme activity and lettuce growth. *Bioresour. Technol.* **2004**, *93*, 21–28. [CrossRef]
19. Bernard, E.; Larkin, R.P.; Tavantzis, S.; Erich, M.S.; Alyokhin, A.; Sewell, G.; Lannan, A.; Gross, S.D. Compost, rapeseed rotation, and biocontrol agents significantly impact soil microbial communities in organic and conventional potato production systems. *Appl. Soil Ecol.* **2012**, *52*, 29–41. [CrossRef]
20. Zhen, Z.; Liu, H.; Wang, N.; Guo, L.; Meng, J.; Ding, N.; Wu, G.; Jiang, G. Effects of manure compost application on soil microbial community diversity and soil microenvironments in a temperate cropland in China. *PLoS ONE* **2014**, *9*, e108555. [CrossRef]
21. Wagg, C.; Bender, S.F.; Widmer, F.; van der Heijden, M.G.A. Soil biodiversity and soil community composition determine ecosystem multifunctionality. *Proc. Natl. Acad. Sci. USA* **2014**, *111*, 5266–5270. [CrossRef]
22. Martínez-Blanco, J.; Lazcano, C.; Christensen, T.H.; Muñoz, P.; Rieradevall, J.; Møller, J.; Antón, A.; Boldrin, A. Compost benefits for agriculture evaluated by life cycle assessment. A review. *Agron. Sustain. Dev.* **2013**, *33*, 721–732. [CrossRef]

23. Yang, W.; Guo, Y.; Wang, X.; Chen, C.; Hu, Y.; Cheng, L.; Gu, S.; Xu, X. Temporal variations of soil microbial community under compost addition in black soil of Northeast China. *Appl. Soil Ecol.* **2017**, *121*, 214–222. [CrossRef]

24. O'Donnell, A.G.; Colvan, S.R. *Biological Diversity and Function in Soils*; Choice Reviews Online; Bardgett, R.D., Usher, M.B., Hopkins, D.W., Eds.; Cambridge University Press: Cambridge, UK, 2005; Volume 43, pp. 43–46, ISBN 0521609879.

25. Van der Heijden, M.G.; Wagg, C. Soil microbial diversity and agro-ecosystem functioning. *Plant Soil* **2013**, *363*, 1–5. [CrossRef]

26. Saccá, M.L.; Caracciolo, A.B.; Di Lenola, M.; Grenni, P. Soil biological communities and ecosystem resilience. In *Soil Biological Communities and Ecosystem Resilience*; Springer: Cham, Switzerland, 2017; pp. 9–24. [CrossRef]

27. Marx, V. Microbiology: The return of culture. *Nat. Methods* **2016**, *14*, 37–40. [CrossRef]

28. Eisenhauer, N.; Schielzeth, H.; Barnes, A.D.; Barry, K.E.; Bonn, A.; Brose, U.; Bruelheide, H.; Buchmann, N.; Buscot, F.; Ebeling, A.; et al. A multitrophic perspective on biodiversity-ecosystem functioning research. *Adv. Ecol. Res.* **2019**, *61*, 1–54. [CrossRef]

29. Ruess, L.; Ferris, H. Decomposition pathways and successional changes. *Nematol. Monogr. Perspect.* **2004**, *2*, 547–556.

30. Ferris, H. Form and function: Metabolic footprints of nematodes in the soil food web. *Eur. J. Soil Biol.* **2010**, *46*, 97–104. [CrossRef]

31. Steel, H.; de la Peña, E.; Fonderie, P.; Willekens, K.; Borgonie, G.; Bert, W. Nematode succession during composting and the potential of the nematode community as an indicator of compost maturity. *Pedobiologia* **2010**, *53*, 181–190. [CrossRef]

32. Ferris, H.; Matute, M.M. Structural and functional succession in the nematode fauna of a soil food web. *Appl. Soil Ecol.* **2003**, *23*, 93–110. [CrossRef]

33. DuPont, S.T.; Culman, S.W.; Ferris, H.; Buckley, D.H.; Glover, J.D. No-tillage conversion of harvested perennial grassland to annual cropland reduces root biomass, decreases active carbon stocks, and impacts soil biota. *Agric. Ecosyst. Environ.* **2010**, *137*, 25–32. [CrossRef]

34. Niles, R.K.; Wall Freckman, D. From the ground up: Nematode ecology in bioassessment and ecosystem health. *Plant Nematode Interact.* **2015**, 65–85. [CrossRef]

35. Kapp, C.; Storey, S.G.; Malan, A.P. Options for soil health measurement in vineyards and deciduous fruit orchards, with special reference to nematodes. *S. Afr. J. Enol. Vitic.* **2013**, *34*, 272–280. [CrossRef]

36. Bulluck, L.R.; Barker, K.R.; Ristaino, J.B. Influences of organic and synthetic soil fertility amendments on nematode trophic groups and community dynamics under tomatoes. *Appl. Soil Ecol.* **2002**, *21*, 233–250. [CrossRef]

37. Hu, C.; Qi, Y. Effect of compost and chemical fertilizer on soil nematode community in a Chinese maize field. *Eur. J. Soil Biol.* **2010**, *46*, 230–236. [CrossRef]

38. Herren, G.L.; Habraken, J.; Waeyenberge, L.; Haegeman, A.; Viaene, N.; Cougnon, M.; Reheul, D.; Steel, H.; Bert, W. Effects of synthetic fertilizer and farm compost on soil nematode community in long-term crop rotation plots: A morphological and metabarcoding approach. *PLoS ONE* **2020**, *15*, e0230153. [CrossRef] [PubMed]

39. Abbott, L.K.; Manning, D.A.C. Soil health and related ecosystem services in organic agriculture. *Sustain. Agric. Res.* **2015**, *4*, 116. [CrossRef]

40. Soil Survey Staff, Natural Resources Conservation Service, United States Department of Agriculture. Web Soil Survey. Available online: https://casoilresource.lawr.ucdavis.edu/gmap (accessed on 21 February 2019).

41. Dufour, R.; Brown, S.; Troxell, D. *Nutrient Management Plan (590) for Organic Systems Western State Implementation Guide*; United States Department, North Carolina Agricultural and Technical State University: Greensboro, NC, USA, 2014; pp. 72–89. [CrossRef]

42. Meyer, R.D. Nitrogen on drip irrigated almonds. In *Years of Discovery. A Compendium of Production and Environmental Research Projects 1972–2003*; Almond Board of California: Modesto, CA, USA, 2004; pp. 284–285.

43. Miranda, K.M.; Espey, M.G.; Wink, D.A. A rapid, simple spectrophotometric method for simultaneous detection of nitrate and nitrite. *Nitric Oxide* **2001**, *5*, 62–71. [CrossRef]

44. Eshel, G.; Levy, G.J.; Mingelgrin, U.; Singer, M.J. Critical evaluation of the use of laser diffraction for particle-size distribution analysis. *Soil Sci. Soc. Am. J.* **2004**, *68*, 736–743. [CrossRef]

45. Culman, S.W.; Snapp, S.S.; Freeman, M.A.; Schipanski, M.E.; Beniston, J.; Lal, R.; Drinkwater, L.E.; Franzluebbers, A.J.; Glover, J.D.; Grandy, A.S.; et al. Permanganate oxidizable carbon reflects a processed soil fraction that is sensitive to management. *Soil Sci. Soc. Am. J.* **2012**, *76*, 494. [CrossRef]

46. Barker, K.R. Nematode extraction and bioassays. In *An Advanced Treatise on Meloidogyne. Volume II: Methodology*; Barker, K.R., Carter, C.C., Sasser, J.N., Eds.; Department of Plant Pathology, North Carolina State University: Raleigh, NC, USA, 1985; pp. 19–35.

47. Bongers, T.; Ferris, H. Nematode community structure as a bioindicator for environmental monitoring. *Trends Ecol. Evol.* **1999**, *14*, 224–228. [CrossRef]

48. Ferris, H.; Bongers, T.; De Goede, R.G.M. A framework for soil food web diagnostics: Extension of the nematode faunal analysis concept. *Appl. Soil Ecol.* **2001**, *18*, 13–29. [CrossRef]

49. Sieriebriennikov, B.; Ferris, H.; de Goede, R.G.M. NINJA: An automated calculation system for nematode-based biological monitoring. *Eur. J. Soil Biol.* **2014**, *61*, 90–93. [CrossRef]

50. Caporaso, J.G.; Lauber, C.L.; Walters, W.A.; Berg-Lyons, D.; Huntley, J.; Fierer, N.; Owens, S.M.; Betley, J.; Fraser, L.; Bauer, M.; et al. Ultra-high-throughput microbial community analysis on the Illumina HiSeq and MiSeq platforms. *ISME J.* **2012**, *6*, 1621–1624. [CrossRef] [PubMed]

51. Parsons, L.S.; Sayre, J.; Ender, C.; Rodrigues, J.L.M.; Barberán, A. Soil microbial communities in restored and unrestored coastal dune ecosystems in California. *Restor. Ecol.* **2020**, *28*, S311–S321. [CrossRef]

52. Callahan, B.J.; McMurdie, P.J.; Holmes, S.P. Exact sequence variants should replace operational taxonomic units in marker-gene data analysis. *ISME J.* **2017**, *11*, 2639–2643. [CrossRef] [PubMed]

53. Magurran, A.E. *Measuring Biological Diversity*; Blackwell Science: Malden, MA, USA, 2004.

54. Louca, S.; Wegener Parfrey, L.; Doebeli, M. Decoupling function and taxonomy in the global ocean microbiome. *Science* **2016**, *353*, 1272–1277. [CrossRef]

55. Bates, D.; Mächler, M.; Bolker, B.M.; Walker, S.C. Fitting linear mixed-effects models using lme4. *arXiv* **2015**, arXiv:1406.5823v1. [CrossRef]

56. Oksanen, A.J.; Blanchet, F.G.; Friendly, M.; Kindt, R.; Legendre, P.; Mcglinn, D.; Minchin, P.R.; Hara, R.B.O.; Simpson, G.L.; Solymos, P.; et al. Vegan: Community Ecology Package. 2019. R Package Version 2.2-0. Available online: http://CRAN.Rproject.org/package=vegan (accessed on 21 August 2019).

57. Wickham, H. Ggplot2. *Wiley Interdiscip. Rev. Comput. Stat.* **2011**, *3*, 180–185. [CrossRef]

58. Mathews, C.R.; Bottrell, D.G.; Brown, M.W. A comparison of conventional and alternative understory management practices for apple production: Multi-trophic effects. *Appl. Soil Ecol.* **2002**, *21*, 221–231. [CrossRef]

59. Forge, T.; Neilsen, G.; Neilsen, D.; Hogue, E.; Faubion, D. Composted dairy manure and alfalfa hay mulch affect soil ecology and early production of "Braeburn" apple on M.9 Rootstock. *HortScience* **2013**, *48*, 645–651. [CrossRef]

60. Sharaf, H.; Thompson, A.A.; Williams, M.A.; Peck, G.M. Compost applications increase bacterial community diversity in the apple rhizosphere. *Soil Sci. Soc. Am. J.* **2021**, *85*, 1–17. [CrossRef]

61. Lazcano, C.; Gómez-Brandón, M.; Revilla, P.; Domínguez, J. Short-term effects of organic and inorganic fertilizers on soil microbial community structure and function: A field study with sweet corn. *Biol. Fertil. Soils* **2013**, *49*, 723–733. [CrossRef]

62. Wittebolle, L.; Marzorati, M.; Clement, L.; Balloi, A.; Daffonchio, D.; Heylen, K.; De Vos, P.; Verstraete, W.; Boon, N. Initial community evenness favours functionality under selective stress. *Nature* **2009**, *458*, 623–626. [CrossRef]

63. Lian, T.; Jin, J.; Wang, G.; Tang, C.; Yu, Z.; Li, Y.; Liu, J.; Zhang, S.; Liu, X. The fate of soybean residue-carbon links to changes of bacterial community composition in Mollisols differing in soil organic carbon. *Soil Biol. Biochem.* **2017**, *109*, 50–58. [CrossRef]

64. Fahrbach, M.; Kuever, J.; Remesch, M.; Huber, B.E.; Kämpfer, P.; Dott, W.; Hollender, J. *Steroidobacter denitrificans* gen. nov., sp. nov., a steroidal hormone-degrading gammaproteobacterium. *Int. J. Syst. Evol. Microbiol.* **2008**, *58*, 2215–2223. [CrossRef]

65. Martínez, S.A.; Dussán, J. Lysinibacillus sphaericus plant growth promoter bacteria and lead phytoremediation enhancer with *Canavalia ensiformis*. *Environ. Prog. Sustain. Energy* **2018**, *37*, 276–282. [CrossRef]

66. Aguirre-Monroy, A.M.; Santana-Martínez, J.C.; Dussán, J. Lysinibacillus sphaericus as a nutrient enhancer during fire-impacted soil replantation. *Appl. Environ. Soil Sci.* **2019**, *2019*, 3075153. [CrossRef]

67. Naureen, Z.; Ur Rehman, N.; Hussain, H.; Hussain, J.; Gilani, S.A.; Al Housni, S.K.; Mabood, F.; Khan, A.L.; Farooq, S.; Abbas, G.; et al. Exploring the potentials of Lysinibacillus sphaericus ZA9 for plant growth promotion and biocontrol activities against phytopathogenic fungi. *Front. Microbiol.* **2017**, *8*, 1477. [CrossRef]

68. Lazicki, P.; Geisseler, D.; Lloyd, M. Nitrogen mineralization from organic amendments is variable but predictable. *J. Environ. Qual.* **2020**, *49*, 483–495. [CrossRef]

69. Horrocks, A.; Curtin, D.; Tregurtha, C.; Meenken, E. Municipal compost as a nutrient source for organic crop production in New Zealand. *Agronomy* **2016**, *6*, 35. [CrossRef]

70. Milkereit, J.; Geisseler, D.; Lazicki, P.; Settles, M.L.; Durbin-Johnson, B.P.; Hodson, A. Interactions between nitrogen availability, bacterial communities, and nematode indicators of soil food web function in response to organic amendments. *Appl. Soil Ecol.* **2021**, *157*, 103767. [CrossRef]

71. Ito, T.; Araki, M.; Komatsuzaki, M.; Kaneko, N.; Ohta, H. Soil nematode community structure affected by tillage systems and cover crop managements in organic soybean production. *Appl. Soil Ecol.* **2015**, *86*, 137–147. [CrossRef]

72. Treonis, A.M.; Austin, E.E.; Buyer, J.S.; Maul, J.E.; Spicer, L.; Zasada, I.A. Effects of organic amendment and tillage on soil microorganisms and microfauna. *Appl. Soil Ecol.* **2010**, *46*, 103–110. [CrossRef]

73. Liu, T.; Yu, L.; Xu, J.; Yan, X.; Li, H.; Whalen, J.K.; Hu, F. Bacterial traits and quality contribute to the diet choice and survival of bacterial-feeding nematodes. *Soil Biol. Biochem.* **2017**, *115*, 467–474. [CrossRef]

74. Venette, R.C.; Ferris, H. Influence of bacterial type and density on population growth of bacterial-feeding nematodes. *Soil Biol. Biochem.* **1998**, *30*, 949–960. [CrossRef]

75. Djigal, D.; Brauman, A.; Diop, T.A.; Chotte, J.L.; Villenave, C. Influence of bacterial-feeding nematodes (Cephalobidae) on soil microbial communities during maize growth. *Soil Biol. Biochem.* **2004**, *36*, 323–331. [CrossRef]

76. Xiao, H.F.; Li, G.; Li, D.M.; Hu, F.; Li, H.X. Effect of different bacterial-feeding nematode species on soil bacterial numbers, activity, and community composition. *Pedosphere* **2014**, *24*, 116–124. [CrossRef]
77. Gebremikael, M.T.; Steel, H.; Buchan, D.; Bert, W.; De Neve, S. Nematodes enhance plant growth and nutrient uptake under C and N-rich conditions. *Sci. Rep.* **2016**, *6*, 32862. [CrossRef] [PubMed]
78. Griffiths, B.S. Microbial-feeding nematodes and protozoa in soil: Their effectson microbial activity and nitrogen mineralization in decomposition hotspots and the rhizosphere. *Plant Soil* **1994**, *164*, 25–33. [CrossRef]
79. Wang, S.; Li, T.; Zheng, Z. Response of soil aggregate-associated microbial and nematode communities to tea plantation age. *Catena* **2018**, *171*, 475–484. [CrossRef]
80. Blanc, C.; Sy, M.; Djigal, D.; Brauman, A.; Normand, P.; Villenave, C. Nutrition on bacteria by bacterial-feeding nematodes and consequences on the structure of soil bacterial community. *Eur. J. Soil Biol.* **2006**, *42*, 70–78. [CrossRef]

 *agronomy*

Article

Comparison of Two Different Management Practices under Organic Farming System

Jiří Antošovský [1], Martin Prudil [2], Milan Gruber [2] and Pavel Ryant [1,*]

1   Faculty of AgriSciences, Mendel University in Brno, Zemědělská 1, 61300 Brno, Czech Republic; jiri.antosovsky@mendelu.cz
2   Central Institute for Supervising and Testing in Agriculture, Hroznová 2, 65606 Brno, Czech Republic; martin.prudil@ukzuz.cz (M.P.); milan.gruber@ukzut.cz (M.G.)
*   Correspondence: pavel.ryant@mendelu.cz

Abstract: Organic farmers usually do not have the opportunity to address the actual symptoms of deficiency through the foliar application of synthetic fertilization, therefore, the main treatment is realized by green manure crop cultivation and application of organic fertilizers. The aim of this long-term experiment was to compare two different production systems with and without livestock in terms of organic farming, and a control variant with no fertilization was also included (treatment 1). The production system without animal husbandry was based on solely the application of renewable external resources (compost or digestate) (treatment 2) and the same fertilization with the addition of auxiliary substances (AS) (treatment 3). The production system with animal husbandry included utilization of fertilizers produced on the farm (fermented urine or manure) using solely farm fertilizers (treatment 4) and in addition with AS (treatment 5). Each treatment had three replications. This work describes the average yields from four experimental years and five experimental localities. Winter wheat, potatoes, winter wheat spelt and legume-cereal mix with corn were used and examined as model crops during the first four years of this long-term research. The highest average yield of winter wheat grain and potato tubers during the first two years of the experiment were obtained after the treatments 2 (7.1 t/ha grain, 33.9 t/ha tubers) and 3 (7.0 t/ha grain, 34.1 t/ha tubers). The several times higher nitrogen content in applied digestate and compost in comparison with fermented urine and manure was probably the reason for such results. On the contrary, the results obtained from the third (spelt) and fourth (LCM and corn) experimental years favored treatment 4 (5.5 t/ha grain, 4.6 cereal unit/ha) and 5 (5.4 t/ha grain, 4.7 cereal unit/ha) from the long-term point of view. After four experimental years, the presented results supported the application of farm fertilizers as a preferable option. The treatments with additional application of AS did not provide a higher yield, therefore, such an application seems unnecessary.

Keywords: organic fertilization; wheat; potatoes; legume-cereal mix; corn; yield

Citation: Antošovský, J.; Prudil, M.; Gruber, M.; Ryant, P. Comparison of Two Different Management Practices under Organic Farming System. *Agronomy* **2021**, *11*, 1466. https://doi.org/10.3390/agronomy11081466

Academic Editors: Mirko Cucina and Luca Regni

Received: 15 July 2021
Accepted: 23 July 2021
Published: 23 July 2021

1. Introduction

The origins of organic farming date back to the first half of the 20th century, although the first law describing organic farming was not published until 1985 in Austria. Organic farming is a precisely defined type of farming with increased emphasis on the environment and its individual components. The main objectives of organic farming are to maintain and improve soil fertility and to utilize as many closed nutrients cycles as possible. The protection of the natural sources in organic farming is possible by prohibiting of mineral nitrogen fertilizers, chemical pesticides, etc. [1]. Organic farmers should minimize the use of non-renewable resources and fossil energy, while preserving biodiversity, natural ecosystems and animal welfare. These measures should ensure sustainable farming. However, the emphasis on animal husbandry is significantly lower in comparison with the past in the Czech Republic, although there has been a partial increase in recent years, especially in cattle breeding [2]. The decrease recorded in animal production can be explained by high acquisition costs of modern equipment suited for today's high standard (stables, milking parlors). Another important reason is the low interest of people working with animals, or in agriculture in general (additional costs for possible robotization). Unfortunately, it is also necessary to mention the low purchase prices of products from animal production competing with cheaper imports, usually with lower quality. Therefore, farmers are looking for another way to earn money. This situation has resulted in narrow crop rotation of economic crops (wheat, barley, oilseed rape, corn) with a declining percentage of improving and fodder crops. Another possibility of economic boost was the construction of biogas stations (more than 500 hundred in CZ, [3]), which resulted in the increase in silage corn production. Corn, with other fodder crops, is used in biogas production instead of animal husbandry. Therefore, the content of quality organic matter in our soils is decreasing, as the 4.5–5 t/ha of organic matter undergoes mineralization during the year. The organic matter in the soil is necessary for optimal soil structure, water-holding capacity, perquisition of humus and nutrient supply. As the numbers of livestock is decreasing, the organic matter is delivered to the soil from biogas stations (digestate) or from compost, as society begins to have higher demands in terms of the sorting and recovery of waste. Therefore, farming without the animal husbandry seems like a possible direction for sustainable farming and it should be examined in comparison with classic organic fertilized produced on farms, especially in conditions of organic farming.

The content of soil nutrients in the Czech Republic is decreasing, especially due to the production export (harvest of main products) and nutrient losses (nutrients leaching or volatilization). On top of that, the basics of a balanced principle are often not respected, as the fertilization with essential macronutrient such as P, K, S are omitted, probably because of the high price of fertilizers. This leads to depleted soil and decreases in crop yields. Fertile soil with an optimal supply of nutrients is, however, necessary for reaching optimal yields [4]. The supply of nutrients in the soil, the content and the quality of organic matter and humus before sowing should be the point of emphasis for organic farmers since it is usually not possible to perform fertilization with mineral fertilizers (especially nitrogen) during vegetation as in conventional agriculture. The balance of n (and other nutrients) in conditions of organic farming is possible to influence by the inclusion of nitrogen-fixing and green manure crops to the crop rotation and by application of organic fertilizers, for example manure, slurry or fermented urine in systems with animal husbandry, compost and digestate in systems without livestock [5]. The correct proportion of crops in the crop ration is enhanced with application of any organic fertilizers are necessary for optimal yield in term of organic farming [6].

A significant amount of short- to medium-term observations were performed to evaluate organic farming. A common issue of these observations was usually only a one-sided focus, for example on soil fertility, crop yield or plant protection. One of the first long-term complex experiments was established by The Research Institute of Organic Agriculture in Switzerland [7,8]. A similar field experiment was founded in the Czech Republic in 2014 by the Central Institute for Supervising and Testing in Agriculture as a result of the situation with animal husbandry listed before. The aim of this long-term experiment is to observe the effect of different systems of production in conditions of organic farming to the soil fertility, plant protection, yield of crops and quality of products. The aforementioned experiment includes seven crops in crop rotation with the idea to repeat the crop rotation at least three times and it was established on five different experimental stations. The presented article is a part of this complex experiment and it describes the average crop yields from the first four experimental years. The aim of this work is to compare two different systems of organic farming—farming based on animal husbandry and farming based on renewable resources without livestock. The basic hypothesis was to confirm that addition of organic matter regardless of the source will increase the crop yields. Another hypothesis worked with the idea, that the application of farm fertilizers results in a higher yield in comparison with renewable external resources. Treatments examined in the experiment also include auxiliary substances as the possible method of influencing yield. Therefore, the last hypothesis for this work was to confirm that the crop yield after this additional application is going to be higher in comparison with the respective variant without this fertilization.

2. Materials and Methods

2.1. Experimental Localities

The experiment was founded as a long-term small-plot field observation using RCBD. The experiment was established on five different experimental stations, therefore in different climate-soil conditions. The main characteristics of each locality are described in Table 1. The presented research started early in 2014 with soil analysis of each experimental location (Table 2). The samples of soil were manually collected by soil sampler from the profile 0–30 cm. The macronutrients content was determined according to the Mehlich III [9]. The content of N_{min} (mineral nitrogen) was determined as a sum of nitrate n (ion-selective electrode—ISE method) and ammonium n (Indolphenol spectrophotometry). After the soil samples were collected, the experimental localities at each station were left as fallow until Autumn. This period serves as a protection against weed. The soil samples were collected from the soil profile 0–30 cm each experimental year after the harvest of model crops to evaluate the content of mineral nitrogen.

Table 1. The description of main climate-soil conditions in the experimental localities.

Experimental Area	Characteristics					
	Meters Above Sea Level	**Main Crop**	**Soil Groups**	**Soil Texture**	**Avg. Precipitation (mm)**	**Avg. Temperature (°C)**
Věrovany	207	sugar beet	Chernozems	clay	502	8.7
Čáslav	260				555	8.9
Jaroměřice nad Rokytnou	425	cereals	Haplic luvisols	clay loam	481	8.0
Horažďovice	475	potatoes	Cambisols	sandy loam	585	7.8
Lípa	505				594	7.5

Table 2. The soil content of macronutrients before sowing (October of 2014).

Experimental Area	Content (mg/kg)						
	P	**K**	**Ca**	**Mg**	**NO_3^-**	**NH_4^+**	**N_{min}**
Věrovany	106	215	3184	136	35.3	0.4	35.7
Čáslav	66	172	3082	160	9.3	0.2	9.5
Jaroměřice nad Rokytnou	90	200	3017	211	8.6	0.5	9.1
Horažďovice	79	143	1711	151	17.6	6.4	24.0
Lípa	69	77	2261	112	12.8	0.8	13.6

2.2. Methodology of the Experiment

The main goal of the presented experiment was to compare the production systems with and without animal husbandry based on the application of different organic fertilizers. There were five different treatments included in the experiment: 1. Unfertilized, 2. Renewable external sources, 3. Renewable external sources + Auxiliary substances (AS), 4. Farm fertilizers and 5. Farm fertilizers + AS. Each treatment was established in three repetitions. Green manure (GM) crops (*Pisum sativum* var. arvense, *Phacelia tanacetifolia*) are also included in the crop rotation. Fertilization during the experimental years is described in Tables 3–5. The doses of farm fertilizers (manure and fermented urine) correspond to their production on a farm with a stock density of 0.8 livestock unit per hectare. Doses of renewable fertilizers (compost and digestate) were adjusted to the same level. The harvested area of each plot was 10 m^2, and respective fertilization was always performed on the 50 m^2 area (protective plot and border-line). The crop rotation examined in the presented study included winter wheat–potatoes–winter wheat spelt–legume-cereal mix/silage corn.

Table 3. Fertilization of winter wheat (2014–2015).

Variant	Fertilization (Dose and Date)		Auxiliary substances [1] (Dose and Date)	
1. Unfertilized				
2. Renewable external sources	Digestate—14 t/ha	April of 2015		
3. Renewable external sources + AS			5 l/ha	May (2x)
4. Farm fertilizers	Fermented urine—14 t/ha			
5. Farm fertilizers + AS			5 l/ha	May (2x)

[1] AS: Mg (MgO)—min 4.0%; K (K$_2$O)—min 1.0%; Z—0.2%; Mn—0.1%; Cu—0.05%; B—0.04%; Fe—0.04%; Mo—0.001% + *Ascophyllum nodosum*.

Table 4. Fertilization of potatoes (2016).

Variant	Fertilization (Dose and Date)				AS [1] (Dose and Date)
1. Unfertilized					
2. Green manure + renewable external sources	Compost—27 t/ha	August of 2015	Digestate—14 t/ha	April of 2016	
3. Green manure + renewable external sources + AS					5 L/ha May (2×)
4. Green manure + farm fertilizers	Manure—27 t/ha		Fermented urine—14 t/ha		
5. Green manure + farm fertilizers + AS					5 L/ha May (2×)

[1] AS: Mg (MgO)—min 4.0%; K (K$_2$O)—min 1.0%; Z—0.2%; Mn—0.1%; Cu—0.05%; B—0.04%; Fe—0.04%; M—0.001%.

Table 5. Fertilization of legume-cereal mix and silage corn (2018).

Variant	Fertilization (Dose and Date)				AS [1] (Dose and Date)
1. Unfertilized					
2. Green manure + renewable external sources	Compost—27 t/ha	August of 2017			
3. Green manure + renewable external sources + AS					0.5 L/ha June
4. Green manure + farm fertilizers	Manure—27 t/ha		Fermented urine—14 t/ha	May of 2018	
5. Green manure + farm fertilizers + AS					0.5 L/ha June

[1] AS: Mg (MgO)—min 4.0%; K (K$_2$O)—min 1.0%; Z—0.2%; Mn—0.1%; Cu—0.05%; B—0.04%; Fe—0.04%; M—0.001%.

The winter wheat (*Triticum aestivum*, cultivar Bohemia) was selected as a first model crop. The sowing of winter wheat was performed in the October of 2014 in the standard sowing rate of 200 kg per hectare. Table 3 describes the organic fertilization during the first experimental year. The two applications of auxiliary substance on leaves were carried out in May. AS was based on mix of natural water-soluble oligopeptide, amino acids, Mg, K, microelements, and extract of seaweed *Ascophyllum nodosum*. The harvest of wheat was carried out at the turn of July and August of 2015. The focal occurrence of weeds was eliminated by the mechanical weeder or by hand (*Galium aparine*). The sporadic occurrence of *Oulema melanopus* and *Puccinia striiformis* was observed without any protection realized. The harvest of spelt was performed by the classic plot harvester.

Potatoes (*Solanum tuberosum*, cultivar Adéla) were cultivated as a followed crop during the second experimental year. The experimental area was fertilized with compost and manure during the August of 2015, after the harvest of winter wheat. The organic fertilizers were incorporated to the soil by stubble cultivator. This operation was followed by sowing of green manure crop (*Pisum sativum* var. arvense). Green manure crop provided an average yield of 2.6 t/ha in dry matter. The cultivated biomass was mulched during the November of 2015, the soil was then prepared by ploughing. The application of organic fertilizers and auxiliary substances during this experimental year is described in Table 4. The spring fertilization of potatoes with organic fertilizers (digestate, fermented urine) was performed early in April of 2016 and it was shortly followed by potatoes planting (3000 kg/ha). The AS was applied on leaves in two separate doses during the May of 2016. The occurrence of *Leptinotarsa decemlineata* and *Phytophthora infestans* was observed during vegetation. The plant protection was realized by the application of allowed products in organic farming (active substance Spinosad and copper oxychloride).

Auxiliary substance applied in the second year was based on mix of natural water-soluble oligopeptide, amino acids, Mg, K, and microelements. The two-phase harvest of potato tubers (mechanically removed from ground and gathered by hands) was carried out in September 2016.

Spelt (*Triticum spelta* L., cultivar Alkor) was chosen as a subsequent crop in the crop rotation after the harvest of potatoes. Sowing of winter wheat spelt was carried out in October 2016 at the standard sowing rate of 200 kg/ha. The organic fertilization of spelt was omitted in this year with the idea to mimic the practices of classic crop production. Therefore, crop rotation with an improving organically fertilized forecrop (potatoes) is usually allowed to omit application of organic fertilizers in the following year (spelt). Another argument to skip the fertilization during this year was a higher habitus and lower resistance level of winter wheat spelt to lodging. The auxiliary substance (bacteria fertilizer for improving non-symbiotic nitrogen fixation) was applied to the soil on variants 3 and 5 before sowing of spelt in the dose of 10 L/ha. Weed control was realized two times during the vegetation by mechanical weeder. The occurrence of *Oulema gallaeciana* and *Blumeria graminis* f. sp. *tritici* was observed during the vegetation. The plant protection was realized by application of sulphur-based fungicide in combination with orange-oil based insecticide. The harvest of spelt was carried out in July of 2017 by classis plot harvester.

The spelt straw with the organic fertilizers were incorporated to the soil by ploughing in August 2017. *Phacelia tanacetifolia* was chosen as a GM crop and it was sown immediately after the organic fertilization (Table 5). The mulching and ploughing were performed in November 2017. The average yield of *Phacelia tanacetifolia* was 3.0 t/ha in dry matter. Crop rotation in the fourth experimental year was divided to meet the specific requirements of the compared production systems. Farming based on animal husbandry has to ensure a sufficient supply of feed, therefore, silage corn (*Zea mays*, cultivar KXB7342) was chosen as a model crop on variants 4 and 5 based on farm fertilizers. Corn was sown at the beginning of April (55 kg/ha) and it was also fertilized by the fermented urine approximately a month after sowing. The legume-cereal mix of spring barley (*Hordeum vulgare*, cultivar Azit) and field pea (*Pisum sativum*, cultivar Eso) harvested for grain was chosen as a model crop for variants 2 and 3 based on farming without animal husbandry. These variants were not fertilized by digestate in spring, because the utilization of digestate by a cultivated mixture would be poor (fertilization by digestate was performed in the next year and fertilization by fermented urine was omitted). The legume-cereal mix was sown at the beginning of April (250 kg/ha 1:1). The application of AS (bacteria fertilizer for improving non-symbiotic nitrogen fixation) was performed in June. The weed control was realized by mechanical weeder in the legume-cereal mix and by hand weeder on corn. The plant protection of legume-cereal mix (*Blumeria graminis*) was realized by the application of sulphur-based fungicide, the protection of corn (*Ostrinia Nubilalis*) was realized by *Trichogramma*. The mixture of barley and pea was harvested at the start of August 2018 by a plot harvester with the additional separation of grain, and silage corn was harvested approximately 14 days later by plot harvester. The yields of legume-cereal mix and corn have been recalculated into the cereal units per hectare for better comparison (cereal unit of field pea = 1 t/h of grain × 1.20; cereal unit of barley = 1 t/ha of grain × 1; cereal unit of silage corn = 1 t/ha of fresh silage × 0.15).

The average content of nitrogen in the applied organic fertilizers during the experimental years is described in Table 6. Every fertilizer was delivered to each experimental area from the same source, except for cattle manure, which was produced individually at each experimental area. The source of digestate, compost and fermented manure differs each year, as it was difficult to obtain organic fertilizers with similar quality.

Table 6. The average n content in applied organic fertilizers during the experiment.

Date	Fertilizer	Content of n (%)
April of 2015—fertilization of winter wheat	Digestate	0.75
	Fermented urine	0.05
August of 2015—fertilization after harvest of wheat	Compost	1.30
	Manure	0.67
April of 2016—spring fertilization of potatoes	Digestate	1.37
	Fermented urine	0.06
August in 2017—fertilization after harvest of spelt	Compost	1.47
	Manure	0.69
May of 2018—fertilization of corn	Fermented urine	0.07%

2.3. Statistical Data Analysis

Statistical evaluation of the monitored parameters was performed by the Statistica 12 CZ program [10]. A two-way analysis of variance (ANOVA, variant of fertilization, experimental locality) and follow-up tests according to Fisher (the LSD test) were conducted. The average crop yield recalculated to cereal unit per hectare over 4 experimental year was analyzed by three-way ANOVA (year, variants of fertilization, locality). The results were expressed as the mean ± standard deviation (SD).

3. Results and Discussion

3.1. First Experimental Year—Winter Wheat

The average grain yield of winter wheat observed after the first experimental year is described in Figure 1. Every treatment provided higher yield compared with the control variant with no fertilization. The highest grain yield (7.1 t/ha) was provided after the sole fertilization with renewable external resources (i.e., digestate). The average grain yield detected on unfertilized control (6.4 t/ha) was lower by almost 11%. Figure 1 also describes the statistical differences between examined variants of fertilization. It is evident from this figure that there were differences between the three main examined treatments—unfertilized, renewable resources and farm fertilizers (both without additional AS). The nitrogen content in farm fertilizers (fermented urine) was significantly lower compared with fertilization based on application of renewable external resources (digestate), as shown in Table 6. This could be possibly the reason for lower yields obtained after the application of farm fertilizers, which is also supported by the content of mineral nitrogen in the soil after winter wheat harvest (Figure 2). It is evident from Figure 2 that the average content of nitrogen in the soil after the harvest was lower (5.18 mg/kg) in the production system based on farm fertilizers in comparison with renewable external resources (6.92 mg/kg). Also, the supply of nutrients (P, K, Ca, Mg) in the soil before the start of the experiment was quite good at each experimental station (Table 2). In addition, the examined fertilization was performed only one time at this point, as this was only the start of the experiment.

Therefore, the differences between examined variants in terms of crop yield were relatively low. A more significant difference in crop yield was expected later in the experiment due to the repeating fertilization, additional cultivation of green manure crops and probably decreased content of soil nutrients. The increase of grain yield after different organic fertilization is also described for example by Šimon et al. [11], and the positive effect of digestate application on wheat is also mentioned by Abubaker et al. [12].

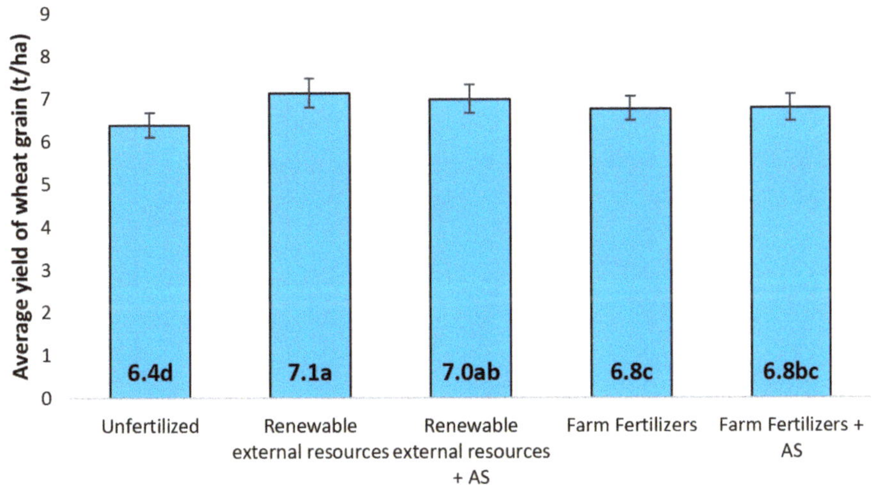

Figure 1. Grain yield of winter wheat (observed treatments, 2015). Error bars represent the standard error (SE). Different letters mean statistical difference ($p \leq 0.5$).

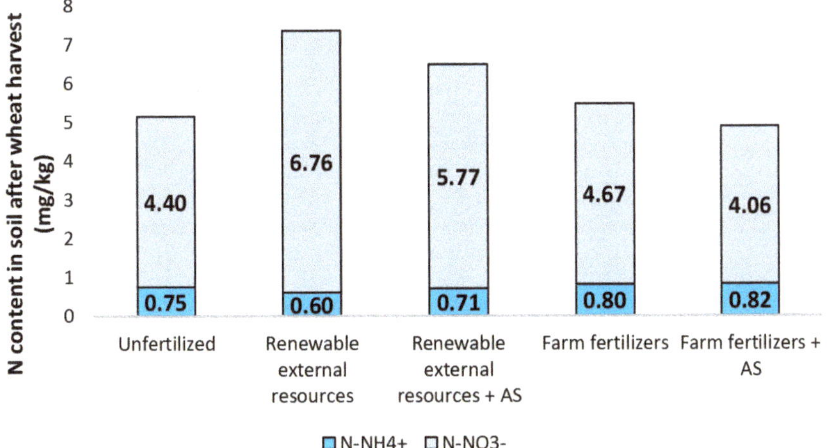

Figure 2. Content of mineral nitrogen in the soil after the harvest of winter wheat (observed treatments, 2015).

The average grain yield of winter wheat in CZ was 6.4 t/ha in the experimental year of 2015 [13]. Each treatment with organic fertilization provided higher yield in comparison with the national average, as described in Figure 1. It necessary to point out that this result is obtained from organic farming, i.e., without application of mineral nitrogen. Therefore, such a result only confirms the importance of organic fertilization to the soil. Rieux et al. [14] are for example also suggesting the utilization of manure for cultivation of cereals as a possible interesting alternative without economic losses to the mineral nitrogen fertilization.

3.2. Second Experimental Year—Potato Tubers

The average yields of potato tubers and the statistical differences between variants of fertilization are described in Figure 3.

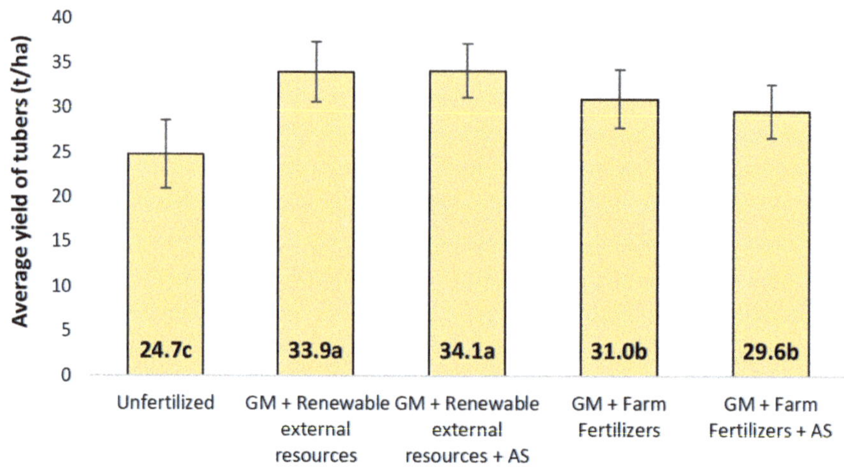

Figure 3. Potato tubers yield (observed treatments, 2016). Error bars represent the standard error (SE). Different letters mean statistical difference ($p \leq 0.5$).

The results are very similar to the result in the first year of the experiment, although the differences between obtained yield were higher. Almost identical yields of tuber were detected on variants based on same production system (with and without animal husbandry). The application of additional auxiliary substances did not provide significantly different yield, which corresponds with the result from previous year with winter wheat. The highest average yield of potato tubers was observed after utilization of renewable external resources (compost + digestate) enhanced with incorporation of green manure crop. The obtained yield (34.1 t/ha) was higher by 38% (9.4 t/ha) compared to the control variant without fertilization and by 9% (3.1 t/ha) in comparison with incorporation of farm fertilizers (manure + fermented urine) enhanced with green manure crop.

The fertilization with manure and, respectively, compost on the soil were carried out in August 2015, and application of fermented urine, respectively digestate was performed in April 2016. Manure was independently produced at each area with a moderate difference in the content of nitrogen (Table 6), while compost was distributed to each area from the same external source. The compost applied in this year of the experiment was characteristic with a very narrow ratio of C/N (9:1). For example, the threshold for commonly used compost described by Loecke et al. [15] should not be lower than 20:1, as a lesser ratio could lead to quick mineralization of organic matter. Therefore, such a low ratio of carbon to nitrogen probably caused a rapid decomposition and nutrient release from compost, as is for example also indicated by Gale et al. [16]. This could be one of the reasons for a higher yield of tubers detected after the treatment with renewable external resources and, therefore, without animal husbandry. For example, Larney et al. [17] and Miller et al. [18] are also recommending fertilization with compost as a preferable option compared to the application of manure, mostly because of the higher content of concentrated nutrients and lower content of carbon, water, and also a lower possibility of nitrogen loss through volatilization. Digestate, respectively fermented urine, were distributed to each area from other external sources as well. Fermented urine characteristically has average content of *n* of about 0.23% [19]. The content of nitrogen in this fertilizer is heavily dependent on the dilution with water, so it can be very variable. In addition, the majority of nitrogen contained in this fertilizer can be characterized as an easily soluble with the highest part in the ammonia *n* form. However, fermented urine applied as fertilizer before potato planting was evaluated as very poor, as the content of nitrogen was only up to 0.06%. Mahimairaja et al. [20] also describe the possible problem with volatilization, e.g., losses of ammonia nitrogen after application of fermented urine. Digestate is also characteristic as a fertilizer with a higher content of mineral nitrogen [21] with a majority in ammonium form [22] and with as much 10 times lower C/N ratio in comparison with farmyard manure [23]. Unfortunately, the digestate applied in this experimental year was characteristic with very high content of nitrogen (1.37%). This could probably be the second reason for achieving a higher yield of potato tuber after treatment without animal husbandry (renewable external resources) in comparison with farm fertilizers. An increase in yield of tubers after the incorporation of digestate or compost as an organic fertilizer was also detected in the research performed by Smatanová [24]. The comparable yields of tubers were also detected in three-year experiment with potatoes based on a comparison of digestate and classic mineral fertilization with UREA [25].

Figure 4 presents the content of mineral nitrogen in the soil after the harvest of potato tubers. It is evident from this figure, that the average content of nitrogen is higher after utilization of farm fertilizers (6.47 mg/kg). This situation supports the previous claim that farming without animal husbandry (application of digestate and compost) positively influenced the crop yield, however, farm fertilizers for crop production are more suitable from the long-term point of view, which was proved next experimental year with the winter wheat spelt without organic fertilization.

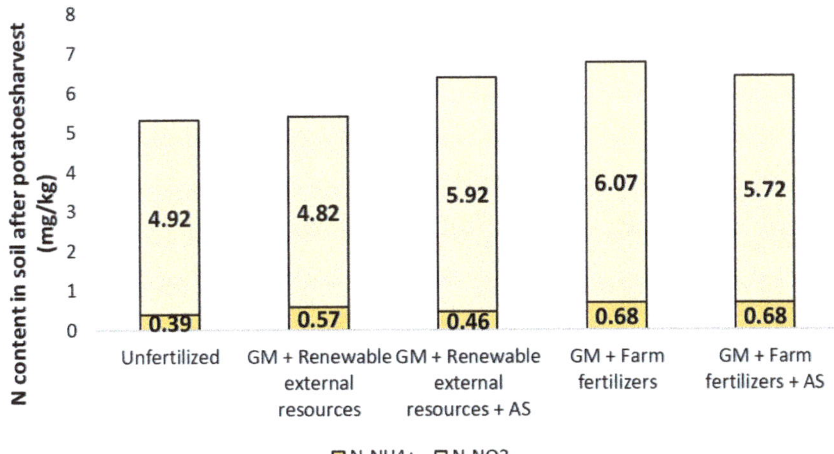

Figure 4. Content of mineral nitrogen in the soil after the harvest of potato tubers (observed treatments, 2016).

The nationwide average yield of potato tubers has reached about 29.4 t/ha in year 2016 [26]. In our research, only the unfertilized control variant provided lower yield. Nevertheless, the average yield of potato tubers detected from this experiment could be again evaluated as very suitable in terms of organic farming as the nationwide average yield of potatoes produced only by organic farmers in the Czech Republic was circa 15 t/ha [2]. El-Sayed et al. [27] or Plaza et al. [28] are recommending organic cultivation of potato tubers as a suitable alternative option in comparison with conventional production without significant reduction of average yield.

3.3. Third Experimental Year—Spelt

The average grain yield of winter wheat spelt and the statistical differences between variants of fertilization are described in Figure 5. Again, the lowest yield of grain (4.6 t/ha) was detected on the control variant without fertilization. By contrast, the highest yield (5.5 t/ha) was provided by fertilization with a farm fertilizer enhanced with a green manure crop. Such result is in contrast in previous experimental years (highest yield of crop after fertilization with renewable resources), which is an interesting development. The highest yield of winter wheat and potatoes detected after the treatment without animal husbandry (renewable external resources) in experimental years 2015–2016 were presumably caused by higher nitrogen content in applied fertilizers (compost, digestate) and narrow C/N ratio. As mentioned in the methodology of the experiment, spelt was not organically fertilized, therefore, the plants were only able to uptake the nutrients from the soil and from the rest of the organic fertilizers applied in previous experimental years. Our result indicates that production systems with animal husbandry, e.g., farm fertilizers (manure, fermented urine), could possibly provide more nutrients to the plants in the following years after application compared to the production systems without livestock, e.g., renewable external resources (compost, digestate). This hypothesis is also confirmed by the content of mineral nitrogen in the soil after the harvest of winter wheat (Figure 6). The average n content in system with farm fertilizers was 4.07 mg/kg, which represents an increase by 0.59 mg/kg in comparison with production systems without animal husbandry (3.48 mg/kg). Comparable result in terms of crop yield was described by several authors [14,16,29,30]. These authors are promoting the fertilization with manure as a preferable option in comparison with compost application. However, it is also possible to find the results describing the application of compost as a more suitable option [17,18,31]. The application of additional auxiliary substances did not provide any significant change in grain yield.

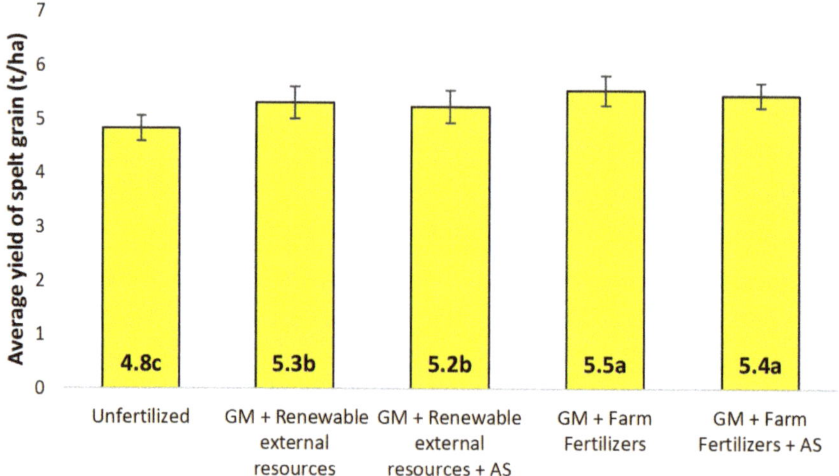

Figure 5. Grain yield of spelt (observed treatments, 2017). Error bars represent the standard error (SE). Different letters mean statistical difference ($p \leq 0.5$).

The average grain yield of winter wheat spelt produced only by organic farming was about 2.7 t/ha in Czech Republic [2,32]. However, the average grain yield of winter wheat spelt detected in this experimental year was about 5.3 t/ha. A result with such a difference could be possibly influenced by crop rotation that improves forecrop two times fertilized by organic fertilizers (potatoes in 2016) and it is again a very good argument for performing any type of organic fertilization and rotating crops. The spelt is also characteristic with possible lodging during vegetation. However, no lodging of plants was observed during the experimental year, which also contributed to the minimum of grain loss during harvest. The spelt also managed to resist disease and pest pressure, which also reduced the possible losses.

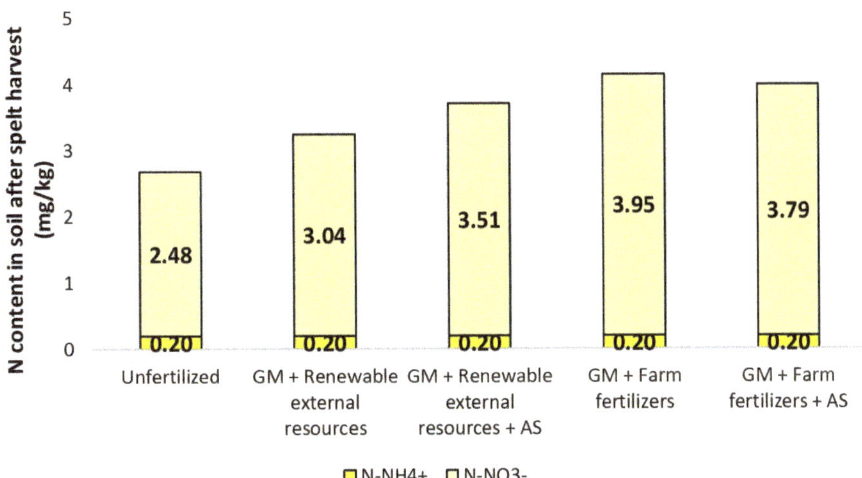

Figure 6. Content of mineral nitrogen in the soil after the harvest of spelt (observed treatments, 2017).

3.4. Fourth Experimental Year—Legume-Cereal Mix and Silage Corn

The average yields of silage corn and legume-cereal mix harvested for grain and their statistical differences are described in the Figure 7. Yields have been recalculated to the cereal units for better comparison. It is evident from these results that the highest yield was provided by the production system with animal husbandry, therefore application of farm fertilizers (manure and fermented urine). There are several possible explanations for this result. Firstly, as this was the fourth experimental year, these results can simply indicate that agriculture based on animal husbandry and therefore application of farm fertilizers could be more suitable in the long-run in terms of slower decomposition and more balanced release of nutrients in comparison with quickly available nutrients (especially *n*) from fertilizers based on renewable external resources. This result is partially supported by two experimental years described by Alburquarque et al. [23], Their results also show that the application of digestate is very beneficial only in the short-term. However, the long-term positive effect of manure fertilization was not proved, as this experiment was performed in horticulture (two crops in one year). Secondly, variants based on farm fertilizers was fertilized two times in this experimental year in comparison with only one application for variants based on renewable external resources (Table 5). This difference in fertilization is caused by different crop rotation and it is explained in the methodology. The final factor to consider is the different crop yield obtained from cultivating of legume-cereal mix and average yield of silage corn (Figure 7). As mentioned before, these yields were recalculated in the cereal units according to corresponding coefficients, but higher average yield of corn in fresh silage (31.5 t/ha) was simply higher in comparison with average grain yield of legume cereal mix (2.2 t/ha of barley + 0.4 t/ha of field pea). An increase in yield of corn or sunflower after organic fertilization is described, for example, by Ahmad et Jabeen [33], Somasundaram et al. [34] or Lehrsch et Kincaid [35]. The yield of spring barley and field pea was also reduced by more pests and diseases in comparison with silage corn.

The content of mineral nitrogen after the harvest is described in Figure 8. The average contents of nitrogen are very similar for both production systems, which may be possibly caused by the difference in crop rotation, as silage corn uptake much more nitrogen to produce adequate biomass in comparison with legume-cereal mix characteristic with lower crop yield.

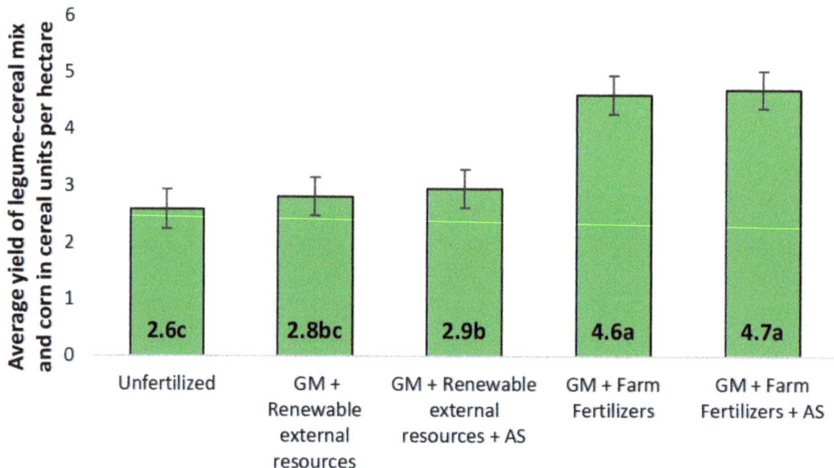

Figure 7. Average yield of legume-cereal mix and silage corn recalculated to the cereal units per hectare (observed treatments, 2018). Error bars represent the standard error (SE). Different letters mean statistical difference ($p \leq 0.5$).

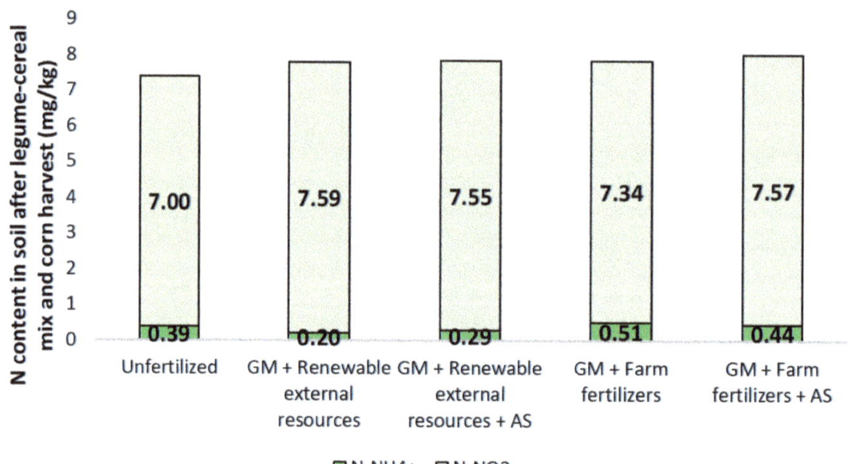

Figure 8. Content of mineral nitrogen after the harvest of legume-cereal mix and corn (observed treatments, 2018).

3.5. Average Yield over Four Experimental Years—Summary

The average yields of crop grain expressed in cereal units per hectare and their statistical differences are described in Figure 9. The highest average yield (6.2 cereal unit per hectare) over four experimental years was observed on the variant with sole application of farm fertilizers. This variant of fertilization provided 24% higher yield compared to the unfertilized control. On a four-year average, both treatments with farm fertilizers achieved slightly higher crop yield compared with both treatments based on production system without animal husbandry. The application of compost and digestate proved to have a short-term positive effect, especially thanks to the significant difference in nitrogen content in applied digestate and fermented urine. Nitrogen rich digestate in combination with compost characteristic with narrow C/N ratio were probably the reason for higher and statistically different yields obtained from first two years of the experiment. Digestate is also a characteristic of an organic fertilizer, but the quality of organic matter, however, is poor as the labile organic matter is used during the fermentation process. This results in narrowing the C/N ratio and high content of organic matter with problematic decomposition [36,37]. By contrast, the crop yield detected in the third and fourth experimental years was in favour for application of farm fertilizers. A possible explanation could be gradual release of nutrients from farm fertilizers (manure and fermented urine) in comparison with quickly available nutrients (especially nitrogen) from renewable external resources (nitrogen rich digestate and compost with narrow C/N ratio). Another possible explanation could be the difference in crop rotation and fertilization in the fourth experimental year, as mentioned before.

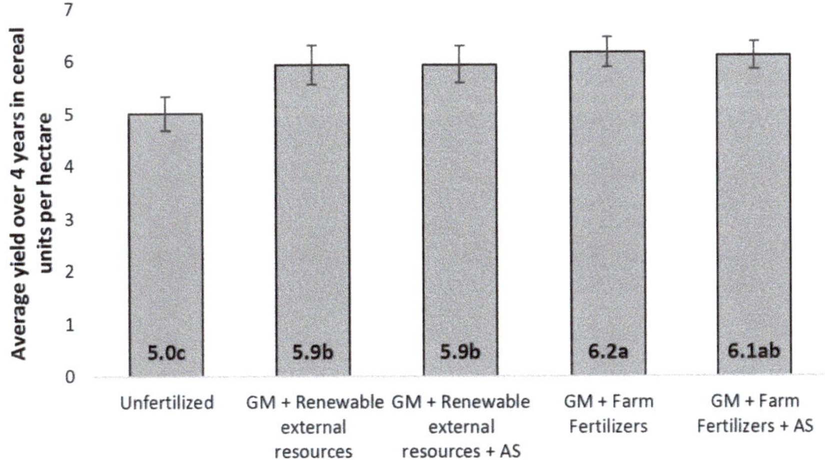

Figure 9. Average crop yield over four experimental years recalculated to the cereal units per hectare units (observed treatments). Error bars represent the standard error (SE). Different letters mean statistical difference. Cereal unit of winter wheat = 1 t/ha of winter wheat grain × 1; cereal unit of potatoes = 1 t/ha of potato tubers × 0.25; cereal unit of winter wheat spelt = 1 t/ha of spelt grain × 1; cereal unit of field pea = 1 t/h of field pea grain × 1.20; cereal unit of spring barley = 1 t/ha of barley grain × 1; cereal unit of silage corn = 1 t/ha of fresh silage × 0.15.

Overall, the average yield achieved during the examined period of the experiment was very high, in some cases even higher in comparison with conventional agriculture. Such results are very promising, and they are only supporting the importance of organic fertilization for maintaining and improving soil fertility, therefore improving possible crop yields. However, it is necessary to point out that the results were obtained from a small plot field experiment with a harvesting area of 10 m^2. The average yield from this area (in kg) was then recalculated to represent the standard agriculture unit of 1 ton per hectare. In common practice, the harvesting area (whole field) is much bigger, which opens a lot of possibilities for biotic and abiotic factors to reduce the crop yield, such as higher occurrence of pest, diseases or weed, higher heterogeneity of the cultivated field, drought and more. It is relatively easier to maintain the small experimental area in comparison with common practice and, therefore, the obtained yield may be influenced by this factor. On the top of that, the cultivated crops were not really damaged by diseases and fungi, probably because of very hot and drought periods during vegetation. The occurrence of insect and weeds were higher, but insects were dealt with the application of allowed products, and weed was removed mechanically or by hand. Before the start of the experiment with organic farming, the examined areas were used for conventional experiments, therefore with intensive agriculture. Overall, the nutrient supply and climate-soil conditions can be evaluated as a very good, which is also a possible explanation for high yields. Another hypothesis behind very high crop yield could be the organic fertilization itself, as the dosage and quality of organic matter in common practice is problematic in our country. Organic fertilization is usually performed mostly by incorporation of straw or green manure crop only. The utilization of digestate has also been rising in recent years, mostly near biogas stations. However, farm fertilizers or compost are being used rather locally, the dosage per hectare can be lower, as the farmers have to distribute their organic fertilizers to more fields.

The application of additional auxiliary substances, therefore fertilizers allowed in organic farming, proved to have statistically insignificant influence on the crop yield in comparison with the same variants without AS. The achieved yield of crop in most of the experimental years was even lower in comparison with same variant without AS, however, the differences were insignificant. Similar results are described by or De Olivera et al. [38]. There is a hypothesis formulated by Holečková [39] that positive results of similar experiments based on application of AS, sometimes referred also as "bioeffectors", are often published only from pot experiments with fully or semi controlled conditions. Several authors [40–42] illustrate a positive effect of similar AS examined in controlled conditions. The inconclusive effect observed in field experiments through the four experimental years could be possibly affected by a much greater competitive relationship between the AS and climate soil conditions in the field in the comparison with pot experiments.

Figure 10 describes the average crop yield detected on five experimental localities over four experimental years. The average yield was recalculated and expressed to the cereal units per hectare. The result corresponds with Tables 1 and 2 respectively (climate-soil and nutrients characteristic of each experimental station). The order in the aforementioned tables is from most optimal to least suitable conditions and it corresponds almost perfectly with the order of each locality in terms of crop yield provided during the experimental period. It is evident, that the average crop yield is decreasing in correlation with increased altitude. The experimental localities Věrovany and Čáslav are also characteristic with the Chernozems soil group with the soil texture clay, which is often considered as the most fertile soil. On the contrary, Horažd'ovice and Lípa can be described as a Cambisols soil group with sandy loam texture.

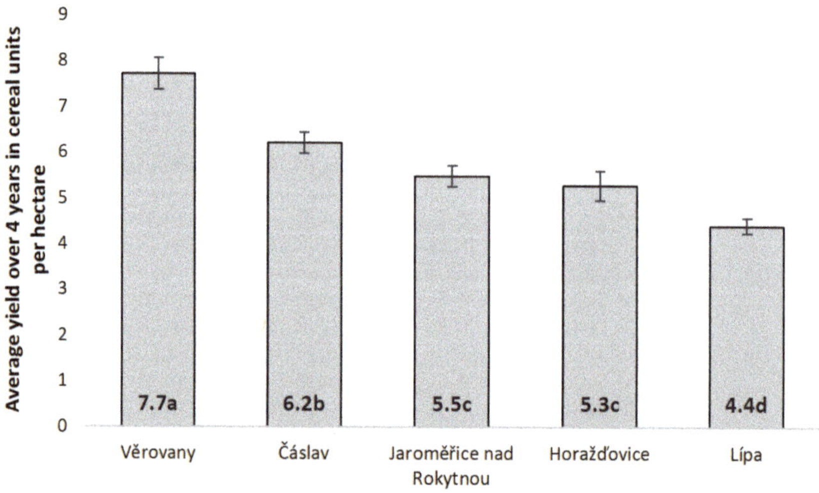

Figure 10. Average crop yield over four experimental years recalculated to the cereal units per hectare (observed experimental localities). Error bars represent the standard error (SE). Different letters mean statistical difference ($p \leq 0.5$).

4. Conclusions

The average crop yields examined over four experimental years indicate the increase in yield of crops after organic fertilization regardless of the production system (with or without animal husbandry) compared to the control variant without fertilization. This result confirms the basic hypothesis given before the start of the experiment, that the addition of organic matter regardless of its source is going to improve the crop yields.

The highest average yields of winter wheat and potato tubers observed after the first and second years of the experiment were provided by the production system without animal husbandry (renewable resources = compost, digestate). The content of quickly available n was higher in these fertilizers compared to the fertilizers with the origin on farms with animal husbandry (manure, fermented urine). Therefore, the observed result is related to the hypothesis, that compost could possibly provide more nutrients to the crops during the first year after application. The average yields observed from the third experimental year with winter wheat spelt was highest in the variant based on farming with animal husbandry, therefore through application of manure or fermented urine. On the contrary, such a result supports the hypothesis that manure in comparison with compost can provide more nutrients to the plants for longer period after fertilization. The result acquired from the last experimental year after recalculating different crop yields (spring barley + field pea and corn) to the cereal units per hectare are also in favour of a production system with animal husbandry. The average crop yields over four experimental years suggest the application of farm fertilizers as the preferable option from the long-term point of view. This was also confirmed by the content of mineral nitrogen in the soil after each experimental year.

Auxiliary substances, fertilizers allowed in organic farming, could possibly enhance the crop yield in organic farming. However, our result obtained over four years did not show statistically differences in average yields of crops observed after addition of auxiliary substance in comparison with the identic variants of fertilization without auxiliary substances. On the contrary, the achieved yield on these variants was insignificantly lower. This result contrasts with the original hypothesis, that the additional fertilization in organic farming is going to result in a higher crop yield.

Author Contributions: Methodology—M.P., M.G.; performing the experiment—M.P., M.G., J.A.; investigation and data analysis—M.P., M.G., J.A.; writing the original draft—J.A.; review and editing—J.A., P.R. All authors have read and agreed to the published version of the manuscript.

Funding: Two experimental years was financially backed by IGA FA MENDELU grants (IP_060/2016 and IP_058/2017).

Data Availability Statement: The data presented in this study are available on request from the corresponding author.

Acknowledgments: Many thanks to the colleagues from the Central Institute for Supervising and Testing in Agriculture for the opportunity to be a part of this long-term work.

Conflicts of Interest: The authors declare no conflict of interest.

References

1. Dvorský, J.; Urban, J. *Basic of Organic Farming According to Council Regulation (ES) nb.834/2007 and Commission Regulation (ES) č.889/2008 with Examples*, 2nd ed.; Central institute for Supervising and Testing in Agriculture: Brno, Czech Republic, 2014; p. 114. (In Czech)
2. Ministry of Agriculture of the Czech Republic. *Yearbook 2018 Organic Farming in the Czech Republic*; MZe: Olomouc, Czech Republic, 2019; p. 80.
3. EBA. *Statistical Report of the European Biogas Association 2018*; European Biogas Association: Brussels, Belgium, December 2018.
4. Marschner, H. *Mineral Nutrition of Higer Plants*, 2nd ed.; Academic Press: London, UK, 2006; p. 672.
5. Urban, J.; Šarapatka, B. *Organic Farming*, 1st ed.; Ministry of Environment in Cooperation with PRO-BIO Association of Organic Farmers: Prague, Czech Republic, 2003; p. 502. (In Czech)
6. Barker, V. *Science and Technology of Organic Farming*; Taylor and Francis Group with CRC Press: Boca Raton, FL, USA, 2010; p. 272.

7.   Maeder, P.; Fliessbach, A.; Dubois, D.; Gunst, L.; Fried, P.; Niggili, U. Soil Fertility and Biodiversity in Organic Farming and Biodiversity in Organic Farming. *Science* **2002**, *296*, 1694–1697. [CrossRef]

8.   Fliess, A.; Oberholzer, H.; Gunst, L.; Maeder, P. Soil organic matter and biological soil quality indicators after 21 years of organic and conventional fading. *Agric. Ecosyst. Environ.* **2007**, *118*, 273–284. [CrossRef]

9.   Zbíral, J.; Malý, D.; Váňa, M. *Analysis of Soil III—Unified Techniques*, 3rd ed.; Central institute for Supervising and Testing in Agriculture: Brno, Czech Republic, 2011; p. 250. (In Czech)

10. StatSoft, Inc. STATISTICA (Data Analysis Software System), Version 12. 2013. Available online: www.statsoft.com (accessed on 5 May 2021).

11. Šimon, T.; Kunzová, E.; Friedlová, M. The effect of digesate, cattle slurry and mineral fertilization on the winter wheat yield and soil quality parameters. *Plant Soil Environ.* **2015**, *61*, 522–527. [CrossRef]

12. Abubaker, J.; Risberg, K.; Pell, M. Biogas residues as fertilisers—Effect on wheat growth and soil microbial activities. *Appl. Energy* **2012**, *99*, 126–134. [CrossRef]

13. Final Harvest Figures—2015. Available online: https://www.czso.cz/csu/czso/final-harvest-figures-2015 (accessed on 10 May 2021).

14. Rieux, C.M.; Vanasse, A.; Chantigny, M.H.; Gelinas, P.; Angers, D.A.; Rochette, P.; Royer, I. Yield and bread-making potential of spring wheat under mineral and organic fertilization. *Crop Sci.* **2013**, *53*, 1139–1147. [CrossRef]

15. Loecke, T.D.; Cambardella, C.A.; Liebman, M. Synchrony of net nitrogen mineralization and maize nitrogen uptake following applications of composted and fresh swine manure in Midwest US. *Nutr. Cycl. Agroecosyst.* **2012**, *93*, 65–74. [CrossRef]

16. Gale, E.; Sullivan, D.M.; Cogger, C.; Bary, A.I.; Hemphill, D.D.; Myhre, E.A. Estimating plant-available nitrogen release from manure, composts, and speciality products. *J. Environ. Qual.* **2006**, *35*, 2321–2332. [CrossRef]

17. Larney, F.J.; Buckey, K.E.; Hao, X.; McGaughey, P.W. Fresh, stockpiled, and composted beef cattle manure: Nutrient levels and mass balance estimates in Alberta and Manitoba. *J. Environ. Qual.* **2006**, *35*, 1844–1854. [CrossRef]

18. Miller, J.J.; Beasley, B.; Drury, C.F.; Zebarth, B.J. Barley yield and nutrient uptake for soil amended with fresh and composted cattle manure. *Agron. J.* **2009**, *101*, 1047–1059. [CrossRef]

19. Škarda, M. *Hospodaření s Organickými Hnojivy*; SZN: Prague, Czech Republic, 1982; p. 324.

20. Mahimaraja, S.; Bolan, N.S.; Hedley, M.J.; MacGregor, M.A. Loses and transformation of nitrogen during composting of poultry manure with different amendments: An incubation experiment. *Bioresour. Technol.* **1994**, *47*, 65–273. [CrossRef]

21. Johansen, A.; Carter, M.S.; Jensen, E.S.; Hauggard-Nielsen, H.; Ambus, P. Effects of digestate from anaerobically digested cattle slurry and plant materials on soil microbial community and emission of CO_2 and N_2O. *Appl. Soil Ecol.* **2013**, *63*, 36–44. [CrossRef]

22. Moller, K.; Muller, T. Effect of anaerobic digestion on digestate nutrient availability and crop growth: A review. *Eng. Life Sci.* **2012**, *12*, 242–257. [CrossRef]

23. Alburquerque, J.A.; de la Fuente, C.; Campoy, M.; Carrasco, L.; Nájera, I.; Baixauli, C.; Caravaca, F.; Roldán, A.; Cegarra, J.; Bernal, M.P. Agricultural use of digestate for horticultural crop production and improvement of soil properties. *Eur. J. Agron.* **2012**, *43*, 119–128. [CrossRef]

24. Smatanová, M. Digestate used as organic fertilizers. *Farmer* **2012**, *18*, 21–22. (In Czech)

25. Ryant, P.; Smatanova, M.; Škarpa, P.; Šimečková, J.; Antosovsky, J.; Hejduk, S. Digestate Utilization As a Fertilizer. In Proceedings of the 25th International Conference on Reasonable Use of Fertilizers, Prague, Czech Republic, 9–11 October 2019; pp. 57–65.

26. Harvest For7ecast—September 2016. Available online: https://www.czso.cz/csu/czso/ari/harvest-forecast-september-2016 (accessed on 10 May 2021).

27. El-Sayed, S.F.; Hassan, H.A.; El-mogy, M.M. Impact of bio- and organic fertilizers on potato yield, quality, and tuber weight loss, after harvest. *Potato Res.* **2015**, *58*, 67–81. [CrossRef]

28. Plaza, A.; Gasiorowska, B.; Makarewicz, A.; Krolikowska, M.A. The yielding of potato fertilized with undersown crops in integrated and organic production system. *J. Plant Breed.* **2013**, *267*, 71–78.

29. Miller, J.J.; Beasley, B.; Drury, C.F.; Zebarth, B.J. Available nitrogen and phosphorus in soil amended with fresh and composted cattle manure. *Can. J. Soil Sci.* **2010**, *90*, 341–354. [CrossRef]

30. Berner, A.; Böhm, H.; Buchecker, K.; Dierauer, H.; Dresow, J.F.; Dreyer, W.; Finckh, M.; Fuchs, A.; Keil, S.; Keiser, A.; et al. *Bio-Kartoffeln: Qualität Mit Jedem Anbau*; Bioland Beratung GmbH: Mainz, Germany, 2010; p. 28.

31. Sanchez, J.E.; Harwood, R.R.; Willson, T.C. Managing soil carbon and nitrogen for productivity and environmental quality. *Agron. J.* **2004**, *96*, 769–775. [CrossRef]

32. Konvalina, P. Winter wheat spelt in organic farming. *Farmer* **2013**, *21*, 35. (In Czech)

33. Ahmad, R.; Jabeen, N. Demonstration of growth improvement in Sunflower (Helianthus annuus) by the use of organic fertilizers under saline conditions. *Pak. J. Bot.* **2009**, *41*, 1373–1384.

34. Somasundaram, E.; Mohamed, A.; Vaipury, K.; Thirukkumaran, K.; Sathyamoorthi, K. Influence ofo rganic sources of nutrients on the yield and economics of crops under maize based cropping system. *J. Appl. Sci. Res.* **2007**, *3*, 1774–1777.

35. Lehrsch, G.A.; Kincaid, A.C. Compost and manure effects of fertilized corn silage yield and nitrogen uptake under irrigation. *Commun. Soil Sci. Plant Anal.* **2007**, *38*, 2131–2147. [CrossRef]

36. Barlog, P.; Hlisnikovsky, L.; Kunzova, E. Yield, content and nutrient uptake by winter wheat and spring barley in response to applications of digestate, cattle slurry and NPK mineral fertilizers. *Arch. Agron. Soil Sci.* **2007**, *66*, 1481–1496. [CrossRef]

37. Gutser, R.; Ebertseder, T.; Weber, A.; Schraml, M.; Schmidhalter, U. Short-term and residual availability of nitrogen afterlong-term application of organic fertilizers on arable land. *J. Plant Nutr. Soil Sci.* **2005**, *168*, 439–446. [CrossRef]
38. De Oliveira, S.M.; Umburanas, R.C.; Pereira, R.G.; De Souza, L.T.; Favarin, J.L. Biostimulants via seed treatment in the promotion of common bean (*Phaseolus vulgaris*) root growth. *Appl. Res. Agrotechnol.* **2017**, *10*, 109–114.
39. Holečková, Z. *The Effect of Bioeffectors on Plant Yield Parameters with a Focus on Increasing Phospohorus Uptake from the Soil*; Czech University of Life Sciences Prague: Prague, Czech Republic, 2018.
40. Backes, C.; Boas, R.L.V.; Santos, A.J.M.; Ribon, A.A.; Bardiviesso, D.M. Foliar application of seaweed extract in potato culture. *Rev. Agric. Neotrop.* **2017**, *4*, 53–57. [CrossRef]
41. Paradiso, R.; Arena, C.; De Micco, V.; Giordano, M.; Aronne, G.; De Pascale, S. Changes in leaf anatomical traits enhanced photosynthetic activity of soybean grown in hydroponics with plant growth-promoting microorganisms. *Front. Plant Sci.* **2017**, *8*, 1–14. [CrossRef]
42. Gravel, V.; Antoun, H.; Tweddell, R.J. Growth stimulation and fruit yield improvement of greenhouse tomato plants by inoculation with *Pseudomonas putida* or *Trichoderma atroviride*: Possible role of indole acetic acid (IAA). *Soil Biol. Biochem.* **2007**, *39*, 1968–1977. [CrossRef]

Article

An Improved Vermicomposting System Provides More Efficient Wastewater Use of Dairy Farms Using *Eisenia fetida*

Xue Liu [1,2,*], Bing Geng [1], Changxiong Zhu [1], Lianfang Li [1] and Frédéric Francis [2,*]

[1] Agricultural Clean Watershed Research Group, Institute of Environment and Sustainable Development in Agriculture, Chinese Academy of Agricultural Science (CAAS), Beijing 100081, China; gengbing@caas.cn (B.G.); zhucx120@caas.cn (C.Z.); lilianfang@caas.cn (L.L.)
[2] Laboratory of Functional and Evolutionary Entomology, Gembloux Agro-Bio Tech, University of Liege, 5030 Gembloux, Belgium
* Correspondence: liuxue@caas.cn (X.L.); frederic.francis@uliege.be (F.F.)

Abstract: Dairy cattle farming produces large amounts of wastewater and it causes environmental pollution and eutrophication of rivers, but the nutrients in the waste could be recycled. Here, an improved vermicomposting system was applied to dairy farm wastewater, and wastewater with a nitrogen content of 100 mg/L and 200mg/L tested with different combinations of organic substrates such as cow manure and rice straw in rural solid waste. Results showed that earthworms could continuously grow, wastewater (N 100mg/L) mixed with rice straw corresponding to the most significant gained weight for *Eisenia fetida* earthworms (2.38 to 9.12-fold), and the earthworms' weight was positively correlated with the C/N ratio, organic matter content, and pH. Compared to the initial state, the system significantly changed physicochemical parameters in nutrients, such as the percentages of total nitrogen, phosphorous, and potassium, which were found to increase in vermicomposting while organic matter content, C/N ratio, and cellulose declined as a function of the vermicomposting period, and the final vermicompost was better for the absorption of plants. These results suggest that continuous wastewater addition improved the effective transformation of organic waste to allow valorizing a broad range of organic residues, and avoid the risk of environmental pollution in dairy cattle farming.

Keywords: wastewater; rice straw; cow manure; *Eisenia fetida*; biological parameters; vermicomposting; waste management; germination; microbial community

Citation: Liu, X.; Geng, B.; Zhu, C.; Li, L.; Francis, F. An Improved Vermicomposting System Provides More Efficient Wastewater Use of Dairy Farms Using *Eisenia fetida*. *Agronomy* **2021**, *11*, 833. https://doi.org/10.3390/agronomy11050833

Academic Editor: Mirko Cucina

Received: 11 March 2021
Accepted: 19 April 2021
Published: 23 April 2021

Publisher's Note: MDPI stays neutral with regard to jurisdictional claims in published maps and institutional affiliations.

1. Introduction

Eutrophication of lakes and rivers is becoming a more serious problem that threatens the safety of human drinking water. Livestock farming constitutes a large proportion of the agricultural non-point source pollutants that cause eutrophication, and China's first pollution-source census demonstrates that dairy cattle farming may be the largest contributor to the problem. Specifically, dairy cattle farming produces 2.2 times more pollutants than non-dairy cattle farming and 15.5 times more than swine farming [1]. The increasing demand for milk from growing human populations requires an innovative solution to mitigate the impact of dairy cattle farms on their surrounding environments.

Backyard breeders and farmers often discard waste directly into surrounding environment, and the current procedures of large-scale operations are costly and somewhat inefficient at removing the organic compounds that contribute to eutrophication. Large-scale livestock farms commonly use anaerobic digestion or aerobic digestion reactors [2]. Up-flow anaerobic sludge blanket (UASB) reactors are routinely used in centralized sewage collection and treatment [3,4], but anaerobic digestion does not remove nitrogen or phosphorus from wastewater. The remaining biogas slurry and residue caused by this secondary pollution is anaerobic, and then inaccessible as crop fertilizer [5]. Activated sludge systems are typically used for aerobic digestion [6,7], but are associated with the production of large amounts of sludge that also needs to be treated. The remaining sludge does not precipitate well due to its high moisture content. Moreover, it has a propensity to expand and float. The use of these two centralized digestion systems has high input costs, high energy consumption, and there are other more actually technical skills for smaller operations.

Using earthworms to treat solid organic waste (i.e., vermicomposting) provides a cheaper and easier alternative to other waste treatment methods [8]. In vermicomposting, worms digest and transform organic waste, also producing a compost to be used in crops for fertilization. Indeed the earthworm metabolic activity during this process enhances the level of nutrients in the transformed wastes [9]. Finally, vermicomposting creates more available nutrients per weight than the original organic substrate before earthworm action, converting organic wastes into excellent bio-fertilizers with abundant humic-like compounds and diverse probiotics that are helpful for agricultural use [10]. Therefore, vermicomposting can facilitate plant growth [11], and unlike pure animal manure, it is an excellent substrate for growing seedlings. Vermicomposting applications correspond to the improvement of conventional composting by adding earthworms to provide a better organic waste stabilization process [12]. Compared to conventional compost, vermicomposting contains higher levels of soluble nutrients and higher quality organic matter [13]. It is a bio-oxidative process in which earthworms and microbes, along with other degradable communities, interact and accelerate the decomposition process of organic waste [14].

Vermicomposting is being utilized within commercially available on-site waste treatment systems; however, there are few reported studies that have examined this medium for the purpose of wastewater treatment [15].

The objective of this study on the improved vermicomposting system was to examine the effects of high nitrogen and carbon contents in the waste liquid from dairy farm wastewater on vermicomposting. The original contributions of this study are that there was a clear difference in evaporation rates of the materials with and without earthworms, and less moisture in the manure added earthworms than in the manure with no earthworms [16]. According to the characteristics of high nitrogen content and high carbon content in the waste liquid of dairy farm wastewater, the continuous addition of wastewater to support the loss of water and maintain the humidity of the system to facilitate the earthworm activities in vermicomposting, in keeping with the sustainable environmental requirements of earthworms' growth. The metabolism of earthworms transforms the nutrients of wastewater, cow dung, and straw into earthworm and earthworm manure with high economic value, achieving the goal of resource utilization and zero release of rural organic wastes. In addition, no after-treatment by-products are generated by achieving the aim of minimizing pollution. Nitrogen loss caused by high temperature composting and air pollution caused by incineration of agricultural organic wastes can be avoided, and this process will not produce after treatment by-products, to achieve the purpose of eliminating pollution.

2. Materials and Methods

2.1. Dairy Farm Waste Collection and Worm Rearing

Rice straw and cow manure were collected from a dairy farm in Yanqing county of Beijing in China; we selected the rice straw harvested in current autumn and fresh manure and wastewater. Young earthworms (*Eisenia fetida*, (Savigny, 1826)) were mass-reared on fresh cow manure in the Institute of Environment and Sustainable Development in Agriculture, Chinese Academy of Agricultural Sciences (CAAS). Fresh cow manure was homogenized and shade-dried in the lab for 2 weeks. Rice straw was cut off to a length of about 1.2 cm. Dairy farm wastewater was diluted with aseptic water and compounded to total nitrogen 100 mg/L and 200 mg/L (pH 7.70). Plastic circular containers with 4 holes in the bottoms were used as vermireactors (13.0 cm × 9.0 cm × 10.0 cm) and the bottoms were sealed with insect netting (mesh size 5 mm).

2.2. Vermicomposting Experimental Set-Up

Two dairy farm waste materials (wastewater and cow manure) and rice straw were used in the experiment. Combinations of three rates of cow manure in rice straw (0, 20 or 40% called M_0, M_{20}, and M_{40} respectively) with two levels of nitrogen content in wastewater (total nitrogen of 100 or 200 mg/L called N_1 and N_2 respectively) corresponding to six treatments were tested (namely M_0N_1, $M_{20}N_1$, $M_{40}N_1$, M_0N_2, $M_{20}N_2$, and $M_{40}N_2$). Each treatment was tested in triplicate, resulting in a total of 126 containers. Twenty non-clitella worms (~275 mg) were selected from the stock rearing and added to each container on the 2nd day. Dairy farm wastewater was added to the system every 3 days to supply the needed amount to avoid desiccation. To measure the digested volume of the wastewater, we measured the added and outflow wastewater from the containers. The moisture content was maintained at 60–80% throughout the vermicomposting period. The containers were kept at 20 °C in the lab. The tests were conducted for 60 days. No rice straw or cow manure was added during this vermicomposting period.

2.3. The Biological Parameters of Earthworms

Over the 60 days of the vermicomposting process, earthworm abundance and biomass were measured every 10 days. Worms were separated by hand, counted, and weighed manually (live weight). Before weighing, worms were washed in distilled water and dried on filter paper. Worms were quickly weighed and returned to their containers to prevent desiccation.

2.4. Physicochemical Parameter Analysis of the Vermicomposting Process and Germination tests

Organic matter (OM), total Kjeldahl nitrogen (TKN), total phosphorus (TP), total potassium (TK), cellulose, pH, and C/N ratio were measured in the initial and the final mixtures [17–19]. The pH was determined in a 1:10 (w/v) aqueous solution (deionized water) with a digital pH meter (Mettler Toledo). The seed germination index was determined at the initial and final points of the improved vermicomposting system using previously described methods [20,21].

2.5. Microbial Diversity Analysis

2.5.1. Sample Collection

R_0 and R_{60} were the same treatments as M_0N_2 at 0 and 60 days, respectively, while CK_0 and CK_{60} were the same treatments as M_0N_2 at 0 and 60 days, respectively, but without earthworms (Table 1).

Table 1. Consumption of wastewater parameters in the improved vermicomposting system until 60th day.

Parameter [a]	Treatments					
	M_0N_1	M_0N_2	$M_{20}N_1$	$M_{40}N_1$	$M_{20}N_2$	$M_{40}N_2$
Volume (mL) [b]	980a	980a	895b	895b	770c	770c
Total nitrogen (mg) [c]	98d	196a	89.5e	89.5e	144c	154b

[a] For each parameter (see text for details), differences among treatments were determined by Tukey's test. Different letters in arrow are significant at $p < 0.05$ (Tukey's test). [b] The total volume of the added wastewater into the improved vermicomposting system because that the dairy farm wastewater was added to the system every 3 days to supply the needed amount to avoid desiccation. [c] The total nitrogen of the added wastewater into the improved vermicomposting system.

2.5.2. DNA Extraction and PCR Amplification

Genomic DNA was extracted from solid samples using the E.Z.N.A. Stool DNA Kit (Omega Bio-Tek, Norcross, GA, USA) according to the manufacturer's instructions. The quality of the extracted DNA was checked by 1% agarose gel electrophoresis and spectrophotometry (optical density at a 260 nm/280 nm ratio). All extracted DNA samples were stored at

−20 °C for further analysis. The V3–V4 hypervariable regions of the 16S rRNA gene were subjected to high-throughput sequencing by Beijing All we gene Tech (Beijing, China) using the Illumina MiSeq PE300 sequencing platform (Illumina, CA, USA). The V3–V4 region of the bacteria 16S rRNA gene were amplified with forward 336F (5′-GTACTCCTACGGGAGGCAGCA-3′) and reverse 806R (5′-GTGGACTACHVGGGTWTCTAAT-3′) universal primers. These primers contain an 8-nucleotide long barcode sequence that is unique to each sample. PCR (Polymerase Chain Reaction) was performed as follows: 95 °C for 5 min, 28 cycles at 95 °C for 45 s, 50 °C for 50 s, and 72 °C for 45 s, with a final extension of 72 °C for 10 min. PCR reactions were performed in triplicate with 25-μL mixtures containing 2.5 μL of 10 × Pyrobest buffer, 2 μL of 2.5 mM dNTPs, 1 μL of each primer (10 μM), 0.4 U of Pyrobest DNA polymerase (TaKaRa), and 15 ng of template DNA. The amplicon mixture was applied to the MiSeq Genome Sequencer (Illumina, San Diego, CA, USA).

2.5.3. Illumina MiSeq Sequencing

Amplicons were extracted from 2% agarose gels and purified using the Axy Prep DNA Gel Extraction Kit (Axygen Biosciences, Union City, CA, USA) according to the manufacturer's instructions and quantified using Quanti Fluor™-ST (Promega, Fitchburg, WI, USA). Purified amplicons were pooled in equimolar concentrations and paired-end sequenced (2 × 300) on an Illumina MiSeq platform according to standard protocols.

2.5.4. Processing of Sequencing Data

The extraction of high-quality sequences was first performed with the QIIME package (v1.2.1; Quantitative Insights Into Microbial Ecology; J Gregory caporaso, University of Colorado, Boulder, CO, USA). Raw sequences were selected based on sequence length, quality, primer, and tag. The raw sequences were selected, and the low-quality sequences were removed based on the following criteria: (i) raw reads were shorter than 110 nucleotides; (ii) 300-bp reads were truncated at any site receiving an average quality score <20 over a 50-bp sliding window (the truncated reads that were shorter than 50 bp were discarded); (iii) exact barcode matching, 2 nucleotide mismatches in primer matching (reads containing ambiguous characters were removed); and (iv) only sequences that overlapped more than 10 bp were assembled according to their overlap sequence (reads that could not be assembled were discarded).

The unique sequence set was classified into operational taxonomic units (OTUs) under the threshold of 97% identity using UCLUST. Chimeric sequences were identified and removed using Usearch (v8.0.1623). The taxonomy of each 16S rRNA gene sequence was analyzed by UCLUST against the Silva119 16S rRNA database using a confidence threshold of 90%.

2.6. Statistical Analysis

All of the reported data were the means of three replicates. One-way analysis of variance (ANOVA) was used to detect significant differences among treatments. Independent Tukey's tests were also performed to evaluate the distribution of the data sets. Statistical analyses were performed using the SPSS statistical package (Version 13.0; SPSS Inc., Chicago, IL, USA). Unless otherwise indicated, $p < 0.05$ was selected to indicate statistical significance.

3. Results and Discussion

3.1. Consumption of Dairy Farm Wastewater during Vermicomposting

After the preliminary experiment, the total nitrogen concentration of the dairy farm wastewater was set at 100 and 200 mg/L. These two concentrations are higher than the standard for pollutant discharge issued by the national livestock breeding industry. In our study, the amount of wastewater input was closed to the amount of water evaporated by the system. The amount of wastewater consumed in the M_0N_1 and M_0N_2 was the largest (Table 1). In these treatments, all of the fillers were straw and explained this result. The crushed stalks were short and rod-shaped, with gaps between them, making it easier for the moisture to drain. Therefore, more wastewater was needed to maintain the system's humidity. In the mixture of other groups that contained cow manure, the latter was crushed into powder, the crevices between pieces of straw were filled, and the water was more easily retained. The rate of water loss was relatively low, the system moisturizing effect was better, and liquid waste did not need to be added as frequently.

In each 100 g improved vermicomposting system, the volume of dairy farm wastewater consumed by M_0N_1 and M_0N_2 was similar. M_0N_2 consumed 196 mg of TKN and M_0N_1 consumed 98 mg. In terms of total nitrogen consumption, M_0N_2 was a more suitable treatment system for dairy farm wastewater. This study provides information on how the nutrient components of wastewater affect the ability of earthworms to grow, thereby absorbing and transforming organic waste into high-value earthworms and earthworm manure. Vermicomposting using these methods discharge no wastewater, achieving the goal of zero pollutants.

3.2. Earthworm Weight Changes during Vermicomposting

After 10 days, the average weights of all earthworms, except for the M_0N_2 conditions, increased (Figure 1). After 20 days, all treatments had continued positive effect on the average earthworm growing. In $M_{20}N_1$, $M_{40}N_1$, $M_{20}N_2$, and $M_{40}N_2$ containing cow manure, significant higher weight gain were observed when compared to earthworms in M_0N_1 and M_0N_2, which had only rice straw. After 30 days, the earthworm's average weight in $M_{20}N_1$, $M_{40}N_1$, $M_{20}N_2$, and $M_{40}N_2$ declined, while the ones in M_0N_1 and M_0N_2 increased. At day 60, the change in the latter was significantly higher than the ones in $M_{20}N_1$, $M_{40}N_1$, $M_{20}N_2$, and $M_{40}N_2$. The average weight change of M_0N_1 was 2.4 times that of $M_{20}N_1$, and 9.1 times that of $M_{40}N_1$. The average weight change of M_0N_2 was 2.1 times that of $M_{20}N_2$ and 1.7 times that of $M_{40}N_2$ (Table 1).

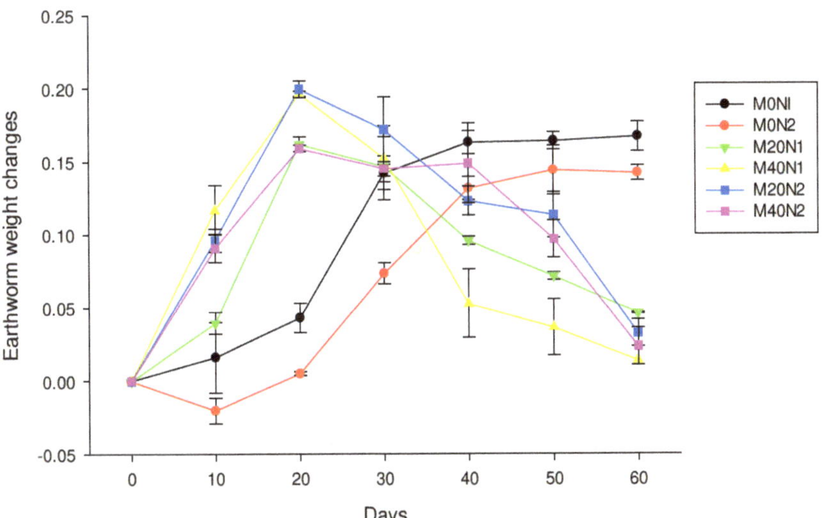

Figure 1. Evolution of *Eisenia fetida* weight according to different proposed diets (in g per worm on 60 day durations).

Earthworms utilize nutrients from different sources to supply their growth needs [22]. The weight change of earthworms was significantly correlated with the C/N ratio.

Dairy wastewater continually supplemented nitrogen, reducing the C/N ratio of all treatments gradually. The M_0N_1 and M_0N_2 conditions became more and more suitable for earthworm growth as this supplementation process continued. High nitrogen can exceed the tolerance limit of earthworms, resulting in slow growth or death [23]. Indeed, at day 20 of the experiment, because of the large amounts of nitrogen in cow manure of $M_{20}N_1$, $M_{40}N_1$, $M_{20}N_2$, and $M_{40}N_2$, the additional dairy wastewater added nitrogen into the mixture, resulting in the increasing nitrogen content, and high nitrogen is toxic to earthworms, so the weight of earthworms began to decrease.

3.3. The Response of Physicochemical Parameters during the Vermicomposting Period

After 60 days, the black color and uniformly disintegrated structure of the mixtures indicated the presence of worm activity [24]. TKN, TP, and TK increased in all of the tests under different treatments after the vermicomposting (Figure 2). The difference between various mixtures was statistically significant (ANOVA, $p < 0.05$) at the end of the process. Traditionally, TKN, TP, TK, organic matter content, and the C/N ratio are considered indicators of decomposition and compost quality [25]. Then, the treatments inoculated with earthworms were more decomposed, and the fillers after vermicomposting were more preferable for bio-organic fertilizer manufacturing [26].

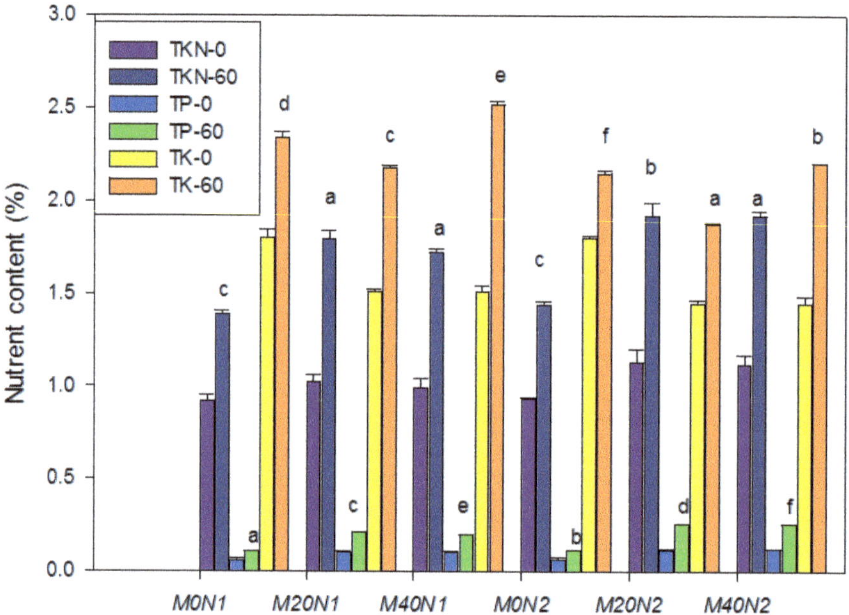

Figure 2. Total Kjeldahl nitrogen (TKN)%, total phosphorus (TP)%, and total potassium (TK)% in different treatments on days 0 and 60 of the experiment. (N-0:TKN% at 0 day; N-60:TKN% at 60 day; TP-0:TP% at 0 day; TP-60:TP% at 60 day; TK-0:TK% at 0 day; TK-60:TK% at 60 day). Compared the difference values of TKN-60 to TKN-0, TP-60 to TP-0, TK-60 to TK-0 in different treatments, the different letters of a, b, c, d were marked on the bar to indicate the significantly different which were determined by Tukey's test ($p < 0.05$).

In the test, there was a 51.1% increase in TKN for M_0N_1, a 75.5% increase for $M_{20}N_1$, and a 69.9% increase for $M_{40}N_1$ compared to the initial level. There was a 54.8% increase in TKN for M_0N_2, a 73.7% increase for $M_{20}N_2$, and a 71.4% increase for $M_{40}N_2$ compared to the initial level (Figure 2). The rate of increase in TKN was higher than 50%, and the mixtures with 20% of cow manure had the highest TKN content. The TKN of 20% cow manure and 40% manure were 1.7% to 1.9% at the 60th day, which were higher than TKN of only-cow manure which was about 1.6% at the 60th day (with *Eisenia fetida*, no wastewater), the TKN of only-straw were 1.3% and 1.4%. TKN of treatments ranged between 1.4% and 1.9% at 60th day, and values were similar to those already reported. The TKN content ranging from 1.2% to 1.7% for vermicompost of paper cattle manure, mill sludge, and dairy sludge were reported [27]. Vermicompost of bilsolids mixed with paper mulch contained

1.7% [28]. But whereas the range reported by values ranging between 1.9% to 3.7% for different types of waste [29]. It indicated that the improved vermicomposting system also had a great impact on nitrogen transformations in the manure, by enhancing nitrogen mineralization, so that mineral nitrogen was retained.

The same trend was observed for TP and TK (Figure 2). There was a 76.4% increase in TP for M_0N_1, a 107.1% increase for $M_{20}N_1$, and a 115.3% increase for $M_{40}N_1$. There was a 73.1% increase in TP for M_0N_2, a 89.1% increase for $M_{20}N_2$, and a 111.3% increase for $M_{40}N_2$. Then, the largest increases in TP which group with 40% cow manure were observed in $M_{40}N_1$ and $M_{40}N_2$ (ANOVA, $p < 0.05$), the lowest increases in TP which group with rice straw-only were observed in M_0N_1 and M_0N_2 (ANOVA, $p < 0.05$). There was a 30.0% increase in TK for M_0N_1, a 43.8% increase for $M_{20}N_1$, and a 29.9% increase for $M_{40}N_1$. There was a 19.5% increase in TK for M_0N_2, a 66.4% increase for $M_{20}N_2$, and a 52.5% increase for $M_{40}N_2$. Therefore, the $M_{20}N_1$ and $M_{20}N_2$ treatments, which contained 20% cow manure, showed the largest increase in TK concentration. Overall, the treatments with earthworms had more nutrients after vermicomposting. During the present investigation, the TP concentration varied from 1.1% to 2.6% and the TK concentration varied from 18.8% to 25.2% at the 60th day, in comparison with recommended standards the TP and TK were found higher in all the samples [30].

Organic matter content decreased in all treatments after vermicomposting (Figure 3A), the difference among mixtures was statistically significant (ANOVA, $p < 0.05$). In comparison to the initial values, there were 5.3%, 19.4%, and 22.1% decreases for M_0N_1, $M_{20}N_1$, and $M_{40}N_1$, respectively, and 8.8%, 15.3%, and 21.6% decreases for M_0N_2, $M_{20}N_2$, and $M_{40}N_2$. The organic matter content varied from 51.3% to 64.4% at the 60th day, which were higher than 45% in the National Organic Fertilizers Standard [31]. The reduction in the organic matter content could be attributed to the rapid decomposition and consumption of the organic materials by earthworms and microorganisms present in their guts [32].

In our experiment, the initial C/N ratio of M_0N_1 and M_0N_2 with only rice straw was significantly higher than that in $M_{20}N_1$, $M_{40}N_1$, $M_{20}N_2$, and $M_{40}N_2$, which also contained cow manure (Figure 3). The C/N ratio significantly decreased among the different mixtures at the end of the experiments (Figure 3B). Decreases of 37.5%, 54.4%, and 54.4% in C/N were observed in M_0N_1, $M_{20}N_1$, and $M_{40}N_1$, and decreases of around 41.4%, 51.2%, and 54.4% in M_0N_2, $M_{20}N_2$, and $M_{40}N_2$. After 60 days, the C/N ratios of M_0N_1 and M_0N_2 were 26.9 and 25.0, the C/N ratios of $M_{20}N_1$ and $M_{20}N_2$ were 17.3 and 19.1, and the C/N ratios of $M_{40}N_1$ and $M_{40}N_2$ were 15.5 and 15.6. Overall the C/N ratios of biosolids in our improved vermicomposting system after treatment were positively correlated with those before. The C/N ratios of the rice straw-only treatment groups were all above 25, in keeping with reports in the literature indicating that a C/N ratio of 25 resulted in the highest and the best fertilizer value of the product [33]. The results of this study revealed that rice straw-only treatments can be used continuously for long time periods and are more capable of eliminating pollutants from dairy farm wastewater.

At the end of the experiment, the cellulose contents of M_0N_1 and M_0N_2 had decreased by 42.1% and 38.3%, respectively, those of $M_{20}N_1$ and $M_{20}N_2$ had decreased by 31.0% and 23.7% respectively, and those of $M_{40}N_1$ and $M_{40}N_2$ had decreased by 15.5% and 26.6% respectively (Figure 3C). This is concordant with the observation that the largest decrease in cellulose content was in the straw-only treatment. Comparing the results of different concentrations of wastewater in the same treatment, cellulose degraded faster in 100 mg/L (TKN) than in 200 mg/L (TKN). This may be because the addition of mild nitrogen which has a lower and mild concentration of nitrogen is more conducive to the metabolism and growth of earthworms, thereby promoting the degradation of cellulose.

The initial pH values of M_0N_1 and M_0N_2, $M_{20}N_1$ and $M_{20}N_2$, and $M_{40}N_1$ and $M_{40}N_2$ were 8.1, 8.6, and 8.9, respectively. At the end of the experiment, the pH values of M_0N_1 and M_0N_2 were 9.0and 9.0, those of $M_{20}N_1$ and $M_{20}N_2$ were 9.1 and 8.9, and those of $M_{40}N_1$ and $M_{40}N_2$ were 9.0 and 9.0, respectively (Figure 3D). The pH of all the treated matrix was about 9.0, and the activity of earthworms changed the pH of the mixture. Vermicomposting can change the pH of different sources of substances through the decomposition of organic matter into inorganic matter [34]. The pH of the vermicomposting process is also strongly correlated with ammonia volatilization [35]. In this study, the continuous supplement of dairy farm wastewater increased the nitrogen content of the improved vermicomposting system, and then the ammonia content due to gaseous losses was replenished. Therefore, the differences in pH resulting from the addition of wastewater were significant over the period of the experiment.

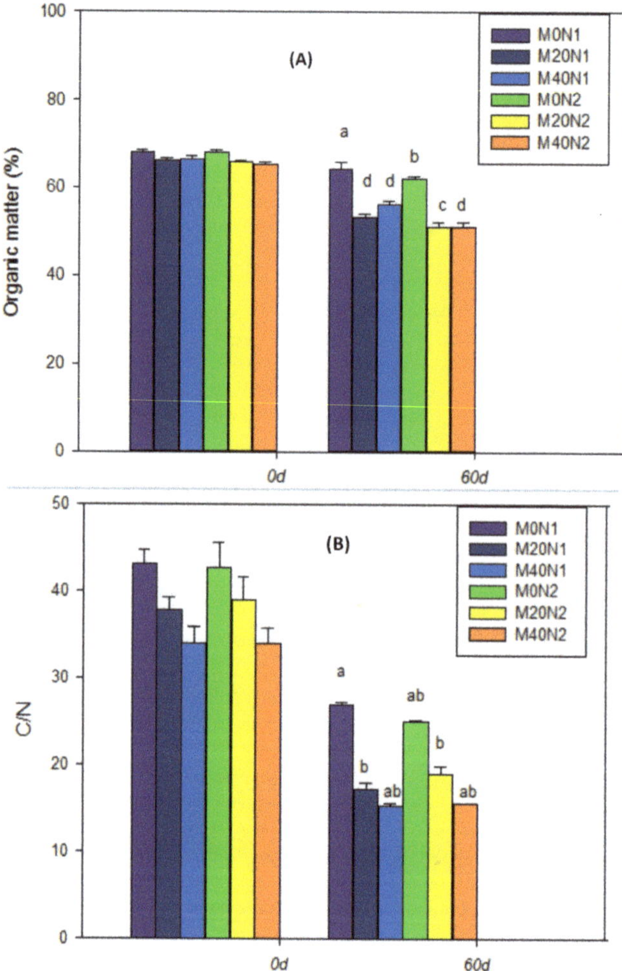

Figure 3. *Cont.*

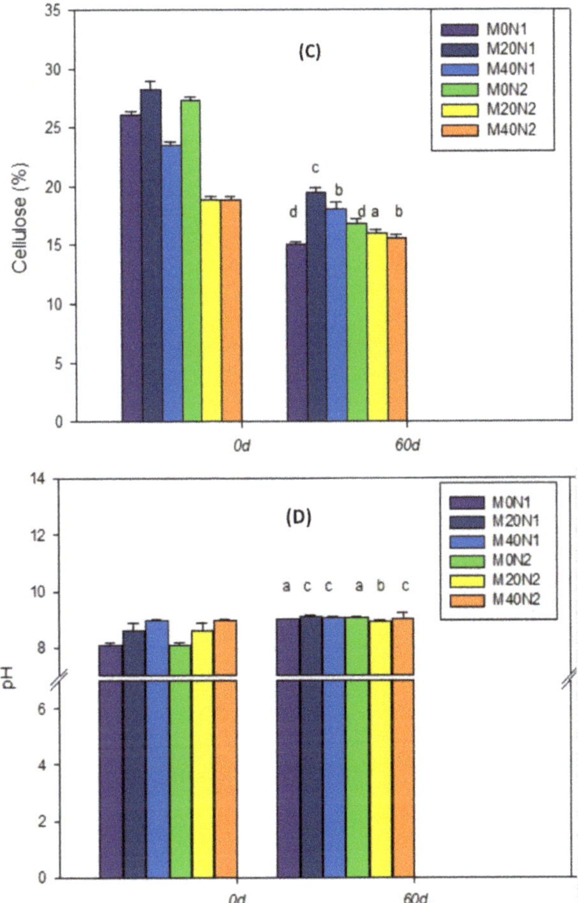

Figure 3. Changes in the physicochemical properties of the different treatments at 0 and 60 days. Comparing the difference values of orgnic matter of 60 day to 0 day (**A**), C/N ratio of 60 day to 0 day (**B**), cellulose of 60 day to 0 day (**C**), pH value of 60 day to 0 day (**D**) in different treatments; different letters in arrows are significant at $p < 0.05$ (Tukey's test).

3.4. Phytotoxicity Test

Immature compost products may have toxic effects on seed germination and seedling growth. We used the germination index to measure the physiological toxicity and maturity of earthworm composting products. Intuitively, when the seed germination coefficient is greater than 50%, the composting is considered to have reached maturity [36]. In the study (Table 2), the germination index of the six tests were all higher than 60%, which demonstrated the improved vermicomposting system form could increase plant productivity. The germination index of M_0N_1 and M_0N_2 were 81% and 81%, respectively, which were significantly higher than those of other treatments. The germination index of $M_{20}N_1$ and $M_{40}N_1$ were 75% and 70%, respectively, and those of $M_{20}N_2$ and $M_{40}N_2$ were 71.86% and 68.66%, respectively. The germination index of these four tests was lower than those of the M_0N_1 and M_0N_2. The M_0N_1 and M_0N_2 groups, which showed the highest germination index, may provide more root growth and better promote the safety of plants [37].

Table 2. Biological parameters of weight changes of earthworms and germination index.

Parameter [a]	Treatments					
	M_0N_1	M_0N_2	$M_{20}N_1$	$M_{40}N_1$	$M_{20}N_2$	$M_{40}N_2$
Weight changes of per worm (g) [b]	$0.16 \pm 0.02a$	$0.11 \pm 0.02ab$	$0.06 \pm 0.02c$	$0.02 \pm 0.003d$	$0.05 \pm 0.05bc$	$0.06 \pm 0.03bc$
Germination index (%)	$81.07 \pm 1.35a$	$81.13 \pm 4.95a$	$75.64 \pm 7.74a$	$70.16 \pm 5.63a$	$71.86 \pm 4.20a$	$68.66 \pm 8.97a$

[a] For each parameter (see text for details), differences among treatments were determined by Tukey's test. Different letters in arrow are significant at $p < 0.05$ (Tukey's test). [b] Weight changes of per worm between day 0 and the 60th day.

3.5. Correlation of Biological and Chemical Indexes

According to the detrended correspondence analysis, the length of the gradient of the first axis was 0.961, indicating that redundancy analysis (RDA) could better explain the relationship between the weight changes of per earthworm and germination index for environment factors compared with canonical correlation analysis. The ordination diagram based on RDA was presented (Figure 4). Similar trends were seen with the M_0N_1 and M_0N_2, which were highly positively correlated with earthworm weight change, germination index, C/N ratio, organic matter, and pH value, but negatively correlated with TKN, TP, TK, and cellulose, which correlated positively with $M_{20}N_1$, $M_{40}N_1$, $M_{20}N_2$, and $M_{40}N_2$. In this study, earthworm weight change was positively correlated with germination index, and the relative seed germination recorded a higher range of earthworm weight gain [38]. The C/N ratio, organic matter, and pH of the mixture had significant effects on the earthworms' growth. The combination of pH value and organic matter had a significant effect on the mixture germination index.

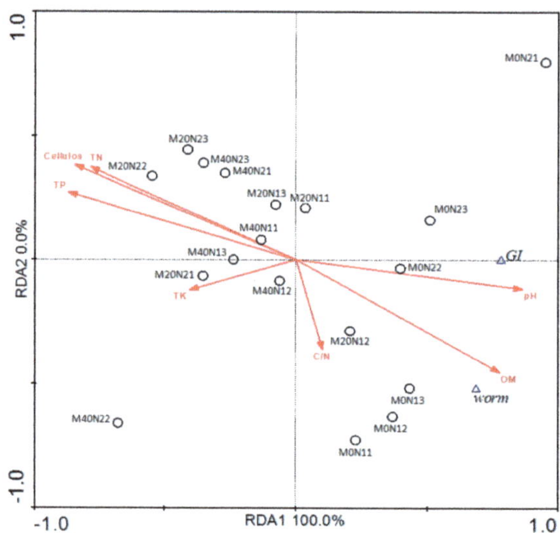

Figure 4. Redundancy analysis of the D_value of the weight of worms and germination rate to the environmental factors in the improved vermicomposting system (after 60 days).

3.6. Composition of the Microbial Community

The relative abundance of bacteria taxonomic groups were presented (Figure 5). Thirty bacteria phyla were identified. At the phylum level, in all groups, at day 0, the dominant phyla were *Proteobacteria*, *Bacteroidetes*, *Actinobacteria*, and *Firmicutes*, which accounted for 99.8% of all bacteria. *Proteobacteria* was the most abundant phylum in all groups, followed by *Bacteroidetes* in groups R_0 and CK_0, indicating that when earthworms were not added, the bacteria in the mixture were mainly *Proteobacteria* and *Bacteroidetes*.

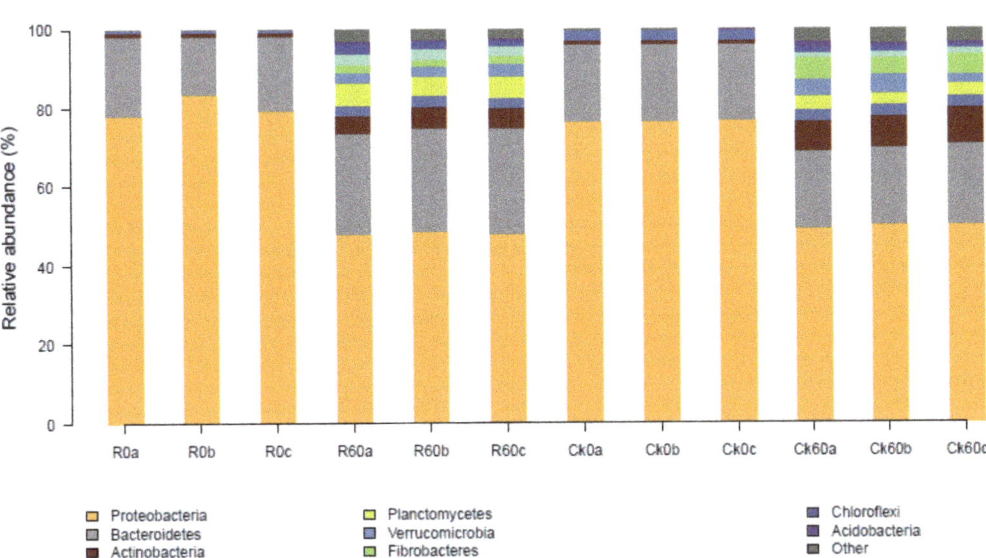

Figure 5. Composition of the bacterial communities at the phylum level. R_{0a}, R_{0b}, and R_{0c}: samples of the rice straw inoculated with earthworms, R_{60a}, R_{60b}, and R_{60c}: samples of the rice straw inoculated with earthworms after 60 days. CK_{0a}, CK_{0b}, and CK_{0c}: samples of the rice straw without earthworms. CK_{60a}, CK_{60b}, and CK_{60c} samples of the rice straw without earthworms after 60 days.

In groups CK and R, at day 60, the dominant phylum in the groups with earthworms were *Proteobacteria* and *Bacteroidetes*, the same as in the groups without earthworms, although the relative abundance of *Proteobacteria* decreased in both treatments, while the abundance of *Bacteroidetes*, *Actinobacteria*, *Firmicutes*, and *Verrucomicrobia* increased. The latter phyla have also been reported to dominate in compost studies without earthworms, which are generally involved in the degradation of organic waste [39]. *Proteobacteria* and *Actinobacteria* are lignin-degrading bacteria [40] while *Bacteroidetes* are thought to degrade cellulose and chitin [41]. *Firmicutes* produce celluloses, lipases, proteases, and other extracellular enzymes that degrade lignin, cellulose, sugars, and amino acids [42]. Finally, *Verrucomicrobia* is mainly found in lakes, drinking water, and other natural freshwater environments [43].

Original composting and vermicomposting modified the original microbial communities of the waste in diverse ways. The differences in microbiota patterns can be attributed to the physicochemical composition of the wastes and the interaction of the earthworms in the microbial communities [44]. Representative bacteria of mature compost (*Actinobacteria*) were more abundant in vermicomposting, and *Bacteroidetes* were more abundant in compost [45]. This resulted in the greater similarity of the microbial community structure between the products of natural placement and maturation via vermicomposting. This indicates that the microbial community composition was affected by the presence of earthworms and by the continuous addition of waste liquid.

4. Conclusions

In summary, the study presented here permits us to develop an improved vermicomposting system, in order to have efficient recycling of wastewater and reuse dairy wastewater, rice straw, and cow manure simultaneously. We demonstrated that earthworms could alter the biochemistry of dairy wastewater and other solid waste and accelerate the stabilization, maturity, and microbial community composition of organic waste. The C/N ratio had the highest effects on earthworm growth, and the increase of earthworm weight promoted the mineralization of solid waste; also, the percentages of total nitrogen, phosphorous, and potassium increased, while organic matter content, C/N ratio, and cellulose declined as a function of the vermicomposting period. For future experiments, we could proceed with the response of electrical conductivity and heavy metals during the vermicomposting period, and the effect of other agricultural organic wastes. This is a feasible method for the simultaneous disposal of organic wastes, especially in poor countries, because it incurs lower costs and has a lower impact on the environment. Our laboratory-scale experiment processing cow manure with earthworms might not fully duplicate large-scale commercial conditions, but provides valuable insights into the process and the changes brought about by earthworm activity.

Author Contributions: Data curation—formal analysis & writing, X.L.; review, F.F.; funding acquisition, C.Z., B.G., L.L.; supervision, F.F., and C.Z. All authors have read and agreed to the published version of the manuscript.

Funding: This work was financially supported by National Water Pollution Control and Treatment Science and Technology Major Project in China (2017ZX07603002, 2015C03004), the National Science Foundation of Beijing (8182059).

Institutional Review Board Statement: Not applicable.

Informed Consent Statement: Not applicable.

Data Availability Statement: The data presented in this study are available on request from the corresponding author.

Acknowledgments: The authors would like to thank the University of Liège-Gembloux Agro-Bio Tech and more specifically Agriculture Is Life for the opportunity platform.

Conflicts of Interest: The authors declare no conflict of interest.

References

1. NDCC. *The First National Census Atlas of Pollution Sources*; China Surveying and Mapping Publishing House: Beijing, China, 2011; ISBN 978750302485. The first National Data Compilation Committee on pollution sources census.
2. Goncalves, M.R.; Costa, J.C.; Marques, I.P.; Alves, M.M. Strategies for lipids and phenolics degradation in the anaerobic treatment of olive mill wastewater. *Water. Res.* **2012**, *46*, 1684–1692. [CrossRef]
3. Haugen, F.; Bakke, R.; Lie, B.; Hovland, J.; Vasdal, K. Optimal design and operation of a UASB reactor for dairy cattle manure. *Comput. Electron. Agr.* **2015**, *111*, 203–213. [CrossRef]
4. Sergey, K.; Vyacheslav, F.; Alla, N. Anaerobic treatment of liquid fraction of hen manure in UASB reactors. *Bioresour. Technol.* **1998**, *65*, 221–225.
5. Daud, M.K.; Rizvi, H.; Akram, M.F.; Ali, S.; Rizwan, M.; Nafees, M.; Jin, Z.S. Review of upflow anaerobic sludge blanket reactor technology: Effect of different parameters and developments for domestic wastewater treatment. *J. Chem.* **2018**, *2018*, 1–13. [CrossRef]
6. Sebastian, B.; Jarosław, D.; Laurence, W. Anaerobic co-digestion of swine and poultry manure with municipal sewage sludge. *Wast. Manag.* **2014**, *34*, 513–521.
7. Zhang, W.Q.; Wei, Q.Y.; Wu, S.B.; Qi, D.D.; Li, W.; Zuo, Z.; Dong, R. Batch anaerobic co-digestion of pig manure with dewatered sewage sludge under mesophilic conditions. *Appl. Energ.* **2014**, *128*, 175–183. [CrossRef]
8. Su, L.L.; Lee, L.H.; Wu, T.Y. Sustainability of using composting and vermicomposting technologies for organic solid waste biotransformation: Recent overview, greenhouse gases emissions and economic analysis. *J. Clean. Prod.* **2016**, *111*, 262–278.
9. Zziwa, A.; Jjagwea, J.; Kizitob, S.; Kabengea, I.; Komakecha, A.J.; Kayondoa, H. Nutrient recovery from pineapple waste through controlled batch and continuous vermicomposting systems. *J. Environ. Manag.* **2021**, *279*, 111784. [CrossRef] [PubMed]
10. Sanchez-Hernandez, J.C.; Domínguez, J. Dual Role of Vermicomposting in Relation to Environmental Pollution. In *Bioremediation of Agricultural Soils*; CRC Press: Boca Raton, FL, USA, 2019; p. 217.
11. Rola, M.; Atiyeh, R.M.; Domínguez, J.; Subler, S.; Edwards, C.A. Changes in biochemical properties of cow manure during processing by earthworms (Eiseniaandrei, Bouché) and the effects on seedling growth. *Pedobiologia* **2000**, *44*, 709–724.
12. Lv, M.; Li, J.; Zhang, W.X.; Zhou, B.; Dai, J.; Zhang, C. Microbial activity was greater in soils added with herb residue vermicompost than chemical fertilizer. *Ecol. Lett.* **2020**, *2*, 209–219. [CrossRef]

13. Sinha, R.K.; Agarwal, S.; Chauhan, K.; Valani, D. The wonders of earthworms & its vermicompost in farm production: Charles Darwin's friends of farmers, with potential to replace destructive chemical fertilizers from agriculture. *Agr. Sci.* **2010**, *1*, 76–97.

14. Patwa, A.; Parde, D.; Dohare, D.; Vijay, R.; Kumar, R. Solid waste characterization and treatment technologies in rural areas: An Indian and international review. *Environ. Technol. Innov.* **2020**, *20*, 101066. [CrossRef]

15. Taylor, M.; Clarke, W.P.; Greenfield, P.F. The treatment of domestic wastewater using small scale vermicompost filter beds. *Ecol. Eng.* **2003**, *21*, 197–203. [CrossRef]

16. Haimi, J.; Huhta, V. Comparison of composts produced from identical wastes by "vermistabilization" and conventional composting. *Pedobiologia* **1987**, *30*, 137–144.

17. Kalembasa, S.J.; Jenkinson, D.S. A Comparative study of titrimetric and gravimetric methods for the determination of organic carbon in soil. *Sci. Food. Agri.* **1973**, *24*, 1085–1090. [CrossRef]

18. Jackson, M.L. *Soil Chemical Analysis*; Prentice Hall of India Pvt. Ltd.: New Delhi, India, 1973.

19. Crosland, A.R.; Zhao, F.J.; McGrath, S.P.; Lane, P.W. Comparison of aqua regia digestion with sodium carbonate fusion for the determination of total phosphorus in soil by inductively coupled plasma atomic emission spectroscopy (ICP). *Commun. Soil Sci. Plan.* **1995**, *26*, 1357–1368. [CrossRef]

20. Awasthi, M.K.; Pandey, A.K.; Khan, J.; Bundela, P.S.; Wong, J.W.C.; Selvam, A. Evaluation of thermophilic fungal consortium for organic municipal solid waste composting. *Bioresour. Technol.* **2014**, *168*, 214–221. [CrossRef]

21. Zhang, L.; Sun, X.Y. Changes in physical, chemical, and microbiological properties during the two-stage co-composting of green waste with spent mushroom compost and biochar. *Bioresour. Technol.* **2014**, *171*, 274–284. [CrossRef]

22. Datta, S.; Singh, J.; Singh, S.; Singh, J. Earthworms, pesticides and sustainable agriculture: A review. *Environ. Sci. Pollut. R.* **2016**, *23*, 8227–8243. [CrossRef]

23. Deka, R.; Kumar, R.; Tamuli, R. *Neurospora crassa* homologue of neuronal calcium sensor-1 has a role in growth, calcium stress tolerance, and ultraviolet survival. *Genetica* **2011**, *139*, 885–894. [CrossRef]

24. Partanen, P.; Hultman, J.; Paulin, L.; Auvinen, P.; Romantschuk, M. Bacterial diversity at different stages of the composting process. *BMC Microbiol.* **2010**, *10*, 1–11. [CrossRef]

25. Gu, W.J.; Zhang, F.B.; Xu, P.Z.; Tang, S.H.; Xie, K.Z.; Huang, X.; Huang, Q.Y. Effects of sulphur and Thiobacillus thioparus on cow manure aerobic composting. *Bioresour. Technol.* **2011**, *102*, 6529–6535. [CrossRef]

26. Li, Y.K.; Yang, X.L.; Gao, W.; Qiu, J.P.; Li, Y.S. Comparative study of vermicomposting of garden waste and cow dung using *Eisenia fetida*. *Environ. Sci. Pollut. Res.* **2020**, *27*, 9646–9657. [CrossRef]

27. Elvira, C.; Sampedro, L.; Benitez, E.; Nogales, R. Bioconversion of solid paper-pulp mill sludge by earthworms. *Bioresour. Technol.* **1998**, *57*, 173–177. [CrossRef]

28. Ndegwa, P.M.; Thompson, S.A. Integrating composting and vermicomposting in the treatment and bioconversion of biosolids. *Bioresour. Technol.* **2001**, *76*, 107–112. [CrossRef]

29. Bernal, M.P.; Paredes, C.; Sanchez-Monedero, M.A.; Cegarra, J. Maturity and stability parameters of composts prepared with a wide range of organic waste. *Bioresour. Technol.* **1998**, *63*, 91–99. [CrossRef]

30. Bordna Mona. *Compost Testing and Analysis Service—Interpretation of Results*; Bordna Mona, Newbridge, Co.: Kildare, Ireland, 2003.

31. NY 525-2012. *People's Republic of China Agricultural Industry Standard*; China Agricultural Publishing House: Beijing, China.

32. El-Haddad, M.E.; Zayed, M.S.; El-Sayed, G.A.M.; Hassanein, M.K.; El-Satar, A.M.A. Evaluation of compost, vermicompost and their teas produced from rice straw as affected by addition of different supplements. *Ann. Ggr. Sci.* **2014**, *59*, 243–251. [CrossRef]

33. Aira, M.; Monroy, F.; Domínguez, J. C to N ratio strongly affects population structure of *Eisenia fetida* in vermicomposting systems. *Eur. J. Soil Biol.* **2006**, *42*, S127–S131. [CrossRef]

34. Garg, P.; Gupta, A.; Satya, S. Vermicomposting of different types of waste using *Eisenia foetida*: A comparatives study. *Bioresour. Technol.* **2006**, *97*, 391–395. [CrossRef]

35. Venkatesh, R.M.; Eevera, T. Mass reduction and recovery of nutrients through vermicomposting of fly ash. *Appl. Ecol. Env. Res.* **2008**, *6*, 77–84. [CrossRef]

36. Zucconi, F.; Pera, A.; Forte, M.; Bertoldi, M.D. Evaluating toxicity of immature compost. *Biocycle* **1981**, *22*, 54–57.

37. Bajsa, O.; Nair, J.; Mathew, K.; Ho, G. Vermiculture as a tool for domestic wastewater management. *Water. Sci. Technol.* **2003**, *48*, 125–132. [CrossRef]

38. Ravindran, B.; Contreras-Ramos, S.M.; Wong, J.W.C.; Selvam, A.; Sekaran, G. Nutrient and enzymatic changes of hydrolysed tannery solid waste treated with epigeic earthworm Eudrilus eugeniae and phytotoxicity assessment on selected commercial crops. *Environ. Sci. Pollut. Res.* **2014**, *21*, 641–651. [CrossRef] [PubMed]

39. Wang, X.J.; Pan, S.Q.; Zhang, Z.Z.; Lin, X.Y.; Zhang, Y.Z.; Chen, S.H. Effects of the feeding ratio of food waste on fed-batch aerobic composting and its microbial community. *Bioresour. Technol.* **2017**, *224*, 397–404. [CrossRef]

40. Bugg, T.D.; Ahmad, M.; Hardiman, E.M.; Singh, R. The emerging role for bacteria in lignin degradation and bio-product formation. *Curr. Opin. Biotech.* **2011**, *22*, 394–400. [CrossRef] [PubMed]

41. Manz, W.; Amann, R.; Ludwig, W.; Vancanneyt, M.; Schleifer, K.-H. Application of a suite of 16S rRNA-specific oligonucleotide probes designed to investigate bacteria of the phylum cytophaga-flavobacter-bacteroides in the natural environment. *Microbiology* **1996**, *142*, 1097–1106. [CrossRef] [PubMed]

42. Lim, J.W.; Chiam, J.A.; Wang, J.Y. Microbial community structure reveals how micro aeration improves fermentation during anaerobic co-digestion of brown water and food waste. *Bioresour. Technol.* **2014**, *171*, 132–138. [CrossRef]

43. Zwart, G.; Crump, B.C.; Kamst-van Agterveld, M.P.; Hagen, F.; Han, S.K. Typical freshwater bacteria: An analysis of avail-able 16S rRNA gene sequences from plankton of lakes and rivers. *Aquat. Microb. Ecol.* **2002**, *28*, 141–155. [CrossRef]
44. Villar, I.; Alves, D.; Pérez-Díaz, D.; Mato, S. Changes in microbial dynamics during vermicomposting of fresh and composted sewage sludge. *Wast. Manag.* **2016**, *48*, 409–417. [CrossRef]
45. Danon, M.; Franke-Whittle, I.H.; Insam, H.; Chen, Y.; Hadar, Y. Molecular analysis of bacterial community succession during prolonged compost curing. *Fems. Microbiol. Ecol.* **2008**, *65*, 133–144. [CrossRef]

agronomy

Article

Compost Tea Induces Growth and Resistance against *Rhizoctonia solani* and *Phytophthora capsici* in Pepper

Ana Isabel González-Hernández, M. Belén Suárez-Fernández, Rodrigo Pérez-Sánchez, María Ángeles Gómez-Sánchez and María Remedios Morales-Corts *

Faculty of Agricultural and Environmental Sciences, University of Salamanca, Avda. Filiberto Villalobos, 119, 37007 Salamanca, Spain; aigonzalez@usal.es (A.I.G.-H.); belensu@usal.es (M.B.S.-F.); rodrigopere@usal.es (R.P.-S.); geles@usal.es (M.Á.G.-S.)
* Correspondence: reme@usal.es

Abstract: Compost teas (CTs) are organic solutions that constitute an interesting option for sustainable agriculture. Those that come from garden waste have been applied in vitro and in vivo on pepper plants to determine its suppressive effect against both *Phytophthora capsici* and *Rhizoctonia solani*. The studied CT showed relevant content in NO_3^-, K_2O, humic acids, and microorganisms such as aerobic bacteria, N-fixing bacteria, and actinobacteria, which play a role in plant growth and resistance. This rich abundance of microbiota in the CT induced a reduction in the relative growth rate of both *P. capsici* and *R. solani* (31.7% and 38.0%, respectively) in in vitro assays compared to control. In addition, CT-irrigated plants displayed increased growth parameters and showed the first open flower one week before those treatments without CTs, which suggests that its application advanced the crop cycle. Concerning pathogen infection, damage caused by both pathogens became more apparent with a one-week inoculation compared to a four-week inoculation, which may indicate that a microbiological and chemical balance had been reached to cope with biotic stresses. Based on these results, we conclude that CT application induces plant growth and defense in pepper plants against *P. capsici* and *R. solani* because of its relevant soluble nutrient content and microbiota richness, which provides a novel point for plant nutrition and protection in horticultural crops.

Keywords: *Capsicum annuun* L.; compost tea; bio-stimulant; plant nutrients; biocontrol; pathogens

Citation: González-Hernández, A.I.; Suárez-Fernández, M.B.; Pérez-Sánchez, R.; Gómez-Sánchez, M.Á.; Morales-Corts, M.R. Compost Tea Induces Growth and Resistance against *Rhizoctonia solani* and *Phytophthora capsici* in Pepper. *Agronomy* **2021**, *11*, 781. https://doi.org/10.3390/agronomy11040781

Academic Editors: Mirko Cucina and Luca Regni

Received: 25 March 2021
Accepted: 14 April 2021
Published: 16 April 2021

Publisher's Note: MDPI stays neutral with regard to jurisdictional claims in published maps and institutional affiliations.

1. Introduction

Large amounts of bio-waste produced every year and should be managed and valorized to provide a well-humified material employable in agriculture. Therefore, recycling organic waste as organic fertilizer is a relevant strategy for sustainable crop production. The application of these fertilizers to the soil could be applied as compost, green manure, farm yard manure, or cereal residues. Some of this biowaste-biosolids, pruning residues, green waste, sewage sludge-displayed good results when properly composted. [1]. This material is obtained through composting, a controlled bioxidative process that requires proper humidity, aeration, and heterogeneous solid organic substrates [2]. Moreover, taking into account the high variability of the parameters that characterize the different composts (pH, electrical conductivity, C-to-N ratio and nutrient content), it is necessary to know their physicochemical and biological properties when applying them to soil or substrate [3]. Nevertheless, compost is a source of macro- and micronutrients not only for plants but also for the microorganisms that support soil health by serving as a quick and easily available source of carbon [4].

Compost from green waste seems to present a lower risk of toxicity than do organic ones, which contain different compounds or microorganisms such as heavy metals, pollutants, viruses, or fecal coliforms [3,5]. In addition, several authors have shown the stimulant and suppressive effect of different compost applications in many cropping systems against several pathogens such as *Fusarium oxysporum*, *Verticillium dahliae*, and *Rhizoctonia solani* [6,7].

Moreover, compost teas (CTs) are organic liquid products that come from the mixture of mature compost with tap water in 1:5 or 1:10 (*v/v*) ratios for a specific period of incubation [8,9]. Several factors affect CT quality, such as the compost type, compost-to-water ratio, and aeration, which altogether modulate the development of beneficial microorganisms [10–12]. Furthermore, CTs are composed of soluble nutrients, and useful compounds and microorganisms (bacteria, actinomycetes, filamentous fungi, oomycetes, and yeasts) that have a synergic effect on suppressing disease and promoting plant growth [13–15]. Soluble nutrients, plant-growth regulators, and humic acids have been previously described as mediators in plant disease suppression [16,17]. Nevertheless, several authors have noted that among the most abundant microorganisms in these extracts are plant-growth promoting rhizobacteria (PGPR), which influence plant growth through different mechanisms (e.g., nitrogen fixation, nutrient solubilization, growth hormones release, and enzymes) that produce a better nutrient availability [18]. These mechanisms seem to be involved in plant-induced resistance, making plants able to cope with subsequent stresses [19]. In short, CTs from green waste seem to display the best suppressive effects against a wide range of different plant disease pathogens [20]. For this reason, the employment of these extracts as a potential eco-friendly alternative to fertilizers and synthetic fungicides has been rising in recent years due to their beneficial effects on crops [20–22], but further research is still required.

Pepper (*Capsicum annuum* L.) is an important horticultural crop worldwide for its nutritional and medicinal value, and has an annual production of more than 38,000,000 tons [23,24]. The pepper fruit is widely consumed for its richness in vitamins, minerals, and nutrients, as well as in phytochemicals like carotenoids, capsaicinoids, flavonoids, ascorbic acid, and tocopherols [25]. Moreover, pepper production could be compromised by different soilborne pathogens such as *Phytophthora capsici* and *Rhizoctonia solani*. *P. capsici*, one of the most economically destructive soil-borne pathogens, causes root rot and foliar, stem and fruit blight of the pepper plant [26]. Moreover, the fungus *R. solani* can also produce seed decay, pre- and post-emergence damping-off, wire stem, leaf decay, root rot, and hypocotyl or taproot with necrotic spots on the pepper [27,28]. The trend towards increasingly organic agriculture has resulted in the need to find new enviro-friendly compounds to control plant diseases. In this context, we hypothesized that the application of a nutrient- and microbial-rich CT obtained from green waste mature compost would improve plant growth and mediate plant defenses. Thus, the main objective of this work was to determine the bio-stimulant and suppressive effect of the extract against *P. capsici* and *R. solani* soil-borne pathogens.

2. Materials and Methods

2.1. Compost Tea Preparation

The compost tea (CT) came from the composting of green and pruning wastes carried out in a garden center located in Salamanca (Spain) (40°57′23″ N; 5°41′8″ W, 775 m a.s.l.). This process was performed in aerated piles of 15 × 2 × 2 m for 180 days. Piles were turned twice per week for eight weeks and once a week during the rest of the bio-oxidative process. Moreover, the moisture of the piles was controlled once a week. The mature compost was obtained under ambient conditions in March. Then, compost was mixed with tap water in a ratio of 1:5 (*v/v*) in polyethylene non-degradable 1000 L containers at room temperature for a brewing period lasting five days. Water had been previously aerated for 8 h to reduce the number of chlorines. This mixture was aerated for five hours every day with a pump. Then, it was filtered with a double-layered cheesecloth, and the aerated CT was stored in a dark container until use.

2.2. Chemical and Microbiological Properties of the Compost Tea

The chemical properties of 9 CT samples were directly analyzed. The pH, electrical conductivity (EC) and C-to-N ratio were determined as described by Morales-Corts et al. [29]. Furthermore, assimilable nutrient contents (NO_3^-, PO_4^{3-}, K_2O, SO_4^{2-}, Ca^{2+} and Mg^{2+}) were analyzed with the nutrient analysis photometer HANNA HI 83225. Finally, humic acids were determined following the alkali-acid method described by Pant et al. [17].

Microbial analysis of the CT was estimated using the serial dilution spread plate method. To determine the microbial population, different selective culture media and CT dilutions were used for microorganism isolation: nutrient agar and 10^{-3} dilution for total aerobic bacteria, Ashby medium, and 10^{-2} dilution for N-fixing bacteria, ISP-2 medium and 10^{-1} dilution for actinobacteria and modified potato dextrose agar medium but no dilution for total fungi and *Trichoderma* ssp. quantification [30–33]. Then, plates were inoculated by depositing on the agar surface 0.1 mL of the CT dilution, which was spread on the media surface with sterile glass beads. Moreover, non-inoculated plates were included as a negative control. Petri dishes were incubated in the dark at 28 °C for 3 to 15 days, depending on the medium. After this time, colony-forming units (CFU) were counted to estimate the cultivable microorganism's population. This experiment was conducted using five replications.

2.3. In Vitro Assays

An initial in vitro test was performed to determine the suppressive effect of the CT against the pathogens *R. solani* and *P. capsici*. The relative growth of both pathogens was measured in Petri dishes of 90 mm in diameter, prepared with sterilized PDA at 60 °C with the CT added after sterilization. Once the medium solidified, a 5-mm diameter PDA plug of *R. solani* or *P. capsici* was inserted into the center of the Petri plates. The CT-to-PDA relation was 1:10. Furthermore, controls of each strain were included in sterilized water and a PDA medium to assess their growth in the absence of the CT. The growth diameter of the pathogens (mm) was assessed at seven days after inoculation. In the end, the RGR was calculated (relative growth rate of the pathogen in comparison to the growth of the colony without the CT) [34]. The essay was carried out including five replications per treatment. Petri dishes were maintained in the dark at 24 °C for 7 days. This experiment was repeated three times. The *P. capsici* strain proceeded from a selection of Neiker (Basque Center of Agricultural Research) *and Rhizoctonia solani* strain from the pathogen collection of the Regional Center for Pest and Diseases Diagnosis, Junta de Castilla y León.

2.4. In Vivo Assays

For in vivo assays, on May 3, pepper seeds (*Capsicum annuum* cv. Morrón de Conserva 4) were germinated in a sterile substrate formed by vermiculite. When the plants had four leaves (30 days after sowing), they were transferred to 15 × 12 × 15 cm pots containing a sterile mix of blond peat (50%), sandy, acid pH, and nutrient-poor soil (37.5%, from Ledesma-Salamanca) and vermiculite (12.5%) as a very nutritive-deficient substrate. The main characteristics of the substrate were: pH 6, 70 mg L^{-1} N, 71 mg L^{-1} P_2O_5, 150 mg L^{-1} K_2O, 35% of organic matter, 0.6 dS m^{-1} EC.

The experiment was carried out in a greenhouse under the following conditions: temperature 24 °C day, 18 °C night and 80% relative humidity. Ten different treatments were considered depending on the CT supply (40 mL was supplied weekly and individually by irrigation), pathogen inoculation, and inoculation time: T1 (CT), T2 (CT + *R. solani* inoculation one week after transplanting), T3 (CT + *P. capsici* inoculation one week after transplanting), T4 (Control plants inoculated with *R. solani* one week after transplanting), T5 (Control plants inoculated with *P. capsici* one week after transplanting), T6 (CT + *R. solani* inoculation four weeks after transplanting), T7 (CT + *P. capsici* inoculation four weeks after transplanting), T8 (Control plants inoculated with *R. solani* four weeks after transplanting), T9 (Control plants inoculated with *P. capsici* four weeks after transplanting) and T10 (Control plants without infection) (Figure 1). For pathogen inoculation, the rice grain method was carried out [35]. Six inoculated rice seeds were placed per pot. Moreover, to determine the effect of CT application, the following plant growth and production parameters were determined: stem diameter measured with a digital caliper, chlorophyll content with a chlorophyll meter Minolta SPAD-502 (both were measured at mid-cycle), dry weight of the root system, and an aerial part (before being weighed, the plants were dried in a P-Selecta-210 oven at 65 °C for 48 h), flowering date, fruit weight, and amount of fruit per plant. The total dry biomass produced per plant was calculated as the sum of the dry fruit, root, and aerial parts. Weights were determined by using a precise balance Sartorius BL150S. Furthermore, the pathogen incidence was also determined following a specific symptom scale (1–5). The experimental design was a randomized complete-block design with eight plants per treatment. This essay was carried out twice during 2017 and 2018.

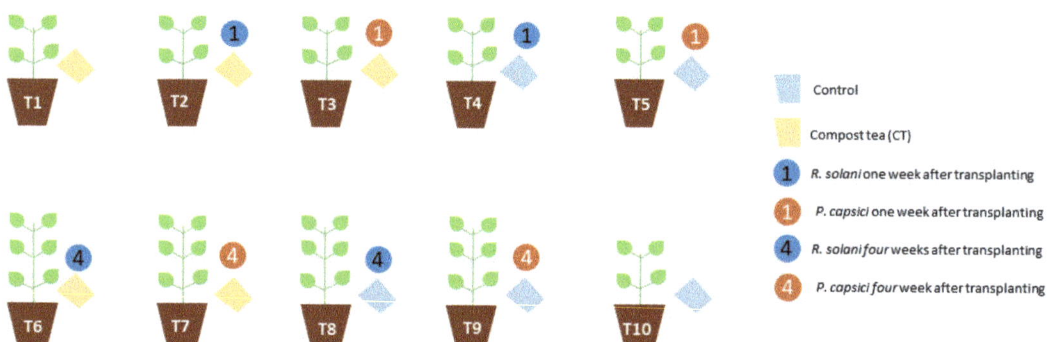

Figure 1. Summary of treatments.

2.5. Statistical Analyses

Statistical analyses were carried out by one-way analysis of variance in Statgraphics Centurion XVIII software (Statistical Graphics Corp., Rockville, MD, USA). Results were presented as means with standard errors, and for comparison the Tukey's Honest Significant Difference (HSD) test with a 95% confidence interval ($p < 0.05$) was used.

3. Results

3.1. Compost Tea Properties

The studied characteristics (pH, EC, C-to-N ratio, assimilable macro and micronutrients, and humic acids) are shown in Table 1. Highlighted are the high NO_3^- and K_2O concentrations (2240.4 and 2851.2 ppm, respectively) and the low C-to-N ratio of the extract, as well as the relevant humic acids amount

Table 1. Chemical characteristics of the studied CT. Results are indicated as mean ± standard deviation.

pH	EC (dS/m)	C/N	NO_3^- (ppm)	P_2O_5 (ppm)	K_2O (ppm)	SO_4^{2-} (ppm)	Ca^{2+} (ppm)	Mg^{2+} (ppm)	Humic Acids (mg/L)
7.16 ± 0.15	1.2 ± 0.14	7.1 ± 0.2	2240.4 ± 225	61.4 ± 25	2851.2 ± 188	43 ± 20	280 ± 17	20 ± 14	198 ± 31

Concerning the microbiological analysis of the CT, the most abundant group of microorganisms was total aerobic bacteria followed by the N-fixing bacteria and actinobacteria (Figure 2). Moreover, *Trichoderma* spp. and fungi were found between 2.7 and 8.7×10^2 cfu/mL, respectively.

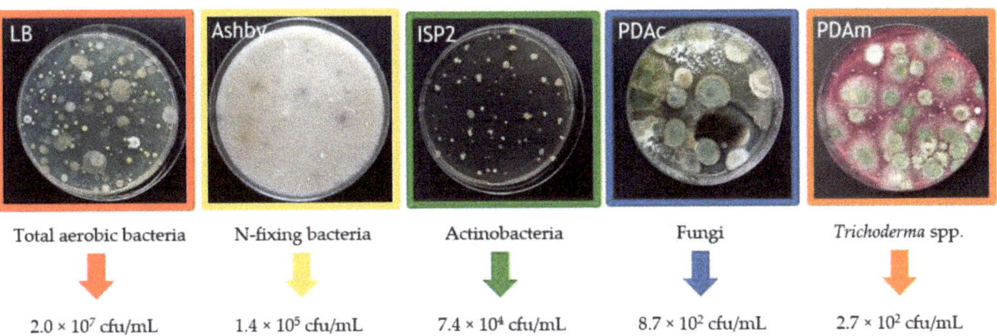

Figure 2. Microbial populations in green waste compost tea.

3.2. Compost Tea Effect in In Vitro Assays against Rhizoctonia solani and Phytophthora capsici

To study in vitro growth of the plant pathogens *R. solani* and *P. capsici* in response to application of the CT, the relative growth rate of each pathogen was calculated. According to this parameter, the relative growth rates of both *R. solani* and *P. capsici* pathogens were 38% and 31.7%, respectively, in comparison to the control treatment (Figure 3 and Table 2).

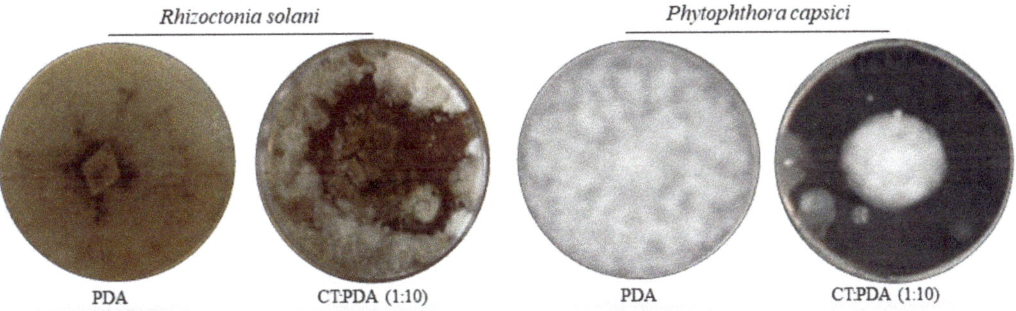

Figure 3. Effect of CT on the relative growth rate of *R. solani* and *P. capsici* in comparison to control treatment.

Table 2. Relative growth rate of *R. solani* and *P. capsici* in CT treatment. Results are indicated as mean ± standard deviation.

Pathogen	RGR (%)
Rhizoctonia solani	38.0 ± 4.2
Phytophthora capsici	31.7 ± 4.6

3.3. Compost Tea Effect in In Vivo Assays against Rhizoctonia solani and Phytophthora capsici

As shown in Table 3, the treatments with compost tea (T1, T2, T3, T6 and T7) advanced the flowering date of pepper plants. CT treatment (T1) also produced an increase in the diameter of the stem in comparison to the control (T10). Likewise, the measurement of this parameter also indicated significant differences in T6 and T7 (CT + *R. solani* or *P. capsici* inoculation four weeks after transplanting) compared to T10. On the other hand, although there were no significant differences, those treatments inoculated with *R. solani* and *P. capsici* one week after transplanting (T4 and T5) showed lower means than the control did for diameter and chlorophyll content parameters.

Table 3. Plant growth parameters of pepper plants grown under the different treatments.

Treatments	Days to Flowering (after Transplanting)	Plant Height (cm)	Stem Diameter (mm)	Chlorophyll Content (SPAD Units)
T1	40.1 a	31.2 abc	8.97 a	37.6 a
T2	39.2 a	33.0 ab	7.72 bc	40.9 a
T3	39.7 a	31.0 abc	9.10 a	38.7 a
T4	48.6 b	26.5 c	7.80 bc	37.1 a
T5	49.0 b	30.0 bc	7.65 c	39.4 a
T6	40.1 a	30.7 abc	9.22 a	39.2 a
T7	41.0 a	36.2 a	8.92 a	38.5 a
T8	48.2 b	33.5 ab	8.60 ab	41.6 a
T9	48.7 b	33.2 ab	8.40 abc	38.6 a
T10	49.2 b	29.7 bc	7.92 bc	39.8 a

Distinct letters in the same column indicate statistically significant differences among treatments as determined by Tukey's honest significant difference (HSD) ($p < 0.05$).

To examine whether the observed changes in growth were related to the CT application and *R. solani* and *P. capsici* diseases, the total amount of fruits, plant dry weight, and pathogen incidence were measured (Table 4). Plants irrigated with the CT (T1) displayed the best results in the total weight of pepper fruit, being two times higher than those of T10. Moreover, CT-treated and -inoculated treatments (T2, T3, T7, and T8) slightly improved the aerial biomass parameters compared to the control (without significant differences). Plants inoculated with *R. solani* or *P. capsici* one week after transplanting (T4 and T5) showed a reduction in shoot and root dry weight, while plants of T5 also presented a decrease in total fruit weight.

Table 4. Fruit production, plant dry weight and *R. solani* and *P. capsici* incidence on pepper plants under CT application.

Treatments	Total Fruit Number	Total Fruit Weight (g)	Mean of Fruit Weight (g)	Root Dry Weight (g)	Shoot Dry Weight (g)	Pathogen Incidence
T1	5	322.19	64.44 a	8.71 ab	5.25 b	0
T2	6	219.25	36.54 bc	7.82 b	5.59 b	0
T3	6	249.58	41.60 b	8.94 ab	6.07 ab	0
T4	4	171.21	42.80 b	4.33 c	4.01 c	2
T5	7	154.44	22.06 d	3.70 cd	3.55 c	3
T6	4	182.07	45.52 b	11.60 a	5.98 ab	0
T7	5	228.38	45.68 b	10.59 a	6.42 a	0
T8	7	209.38	29.91 c	6.51 bc	4.98 b	1
T9	4	147.13	36.78 bc	4.98 c	4.55 bc	2
T10	6	162.23	27.04 c	8.66 ab	4.95 b	0

Distinct letters in the same column indicate statistically significant differences among treatments as determined by Tukey's honest significant difference (HSD) ($p < 0.05$).

Concerning pathogen incidence, it should be pointed out that collapse only occurred in one plant of the T5 treatment. Furthermore, most of the treatment plants showed different symptoms such as black roots, narrow neck, and a thin stem (grade 3 attack severity) and those plants of T4 displayed a grade 2 pathogen attack. Moreover, T8 and T9 (*R.solani* and *P.capsici* inoculated 4 weeks after transplanting) generated less damage than inoculations a week after transplanting (T4 and T5) considering the damage severity of grade 1 and 2, respectively. In addition, it should be noted that there were significant differences among the inoculated *P. capsici* plants one week after transplanting under CT and control conditions (T3 and T5, respectively). Furthermore, there were no differences between CT-treated and inoculated with *R. solani* plants one week after transplanting (T2) and those treated and inoculated four weeks after transplanting (T6), while there were significant differences between T2 and T4.

The total dry biomass produced per plant and treatment are shown in Figure 4. Treatments with the CT (T1, T2, T3, T6, and T7) increased the total biomass of plants. This effect was significant when CT (T1) and control without CT application (T10), were compared. The CT application produced an 36% increase in biomass over control. Plants inoculated with *P. capsici* one week after transplanting (T5) displayed a 50% reduction in biomass production with respect to those inoculated and treated with the CT (T3).

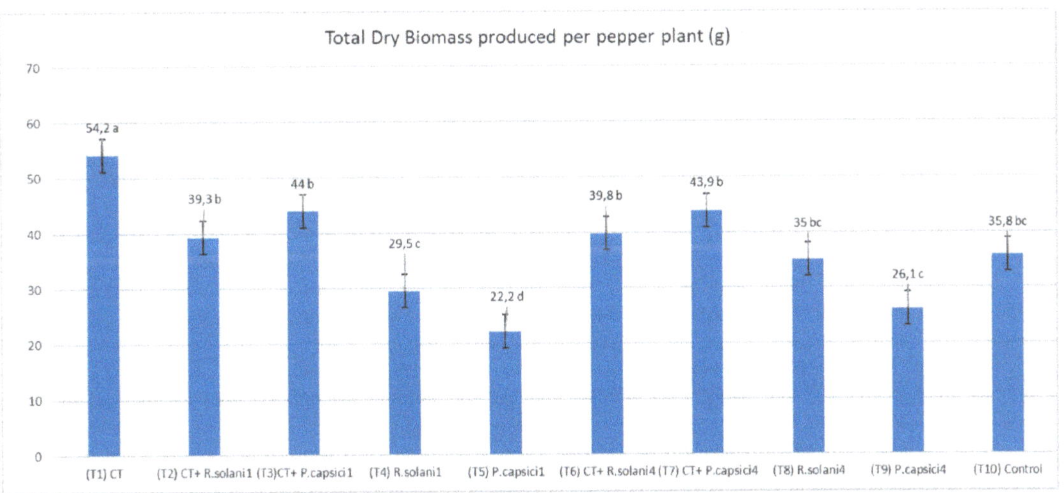

Figure 4. Total dry biomass produced per pepper plant and treatment. T1: CT (compost tea); T2 and T3: CT + pathogen inoculated one week after transplanting; T4 and T5: pathogen inoculated one week after transplanting; T6 and T7: CT + pathogen inoculated four weeks after transplanting; T8 and T9: pathogen inoculated four weeks after transplanting; T10: control.

4. Discussion

The present work showed that green waste CT not only promotes plant development, but also protects plants when exposed to *R. solani* and *P. capsici* pathogens at different times from transplanting.

In this study, a CT coming from green waste was tested. First, the chemical properties of this extract were studied, and the results showed that not only did the CT dose play a relevant role in plant growth and plant resistance, but it also improved soil quality [36,37]. The pH and EC were significantly lower than those obtained by other authors [9,38]. The result of the C-to-N ratio showed a reduction with respect to the original compost (11.4) [29], which indicated a low C amount in relation to the N extracted in the CT. This low ratio points to the extract's high stability and conservation. Nutrient analyses revealed that nitrogen (N) and potassium (K) were well presented in this CT, making teas an interesting source of fertilizer for crop application. Furthermore, high K concentration could also be beneficial for plant resistance against both *R. solani* and *P. capsici* infections, since Amtmann et al. [39] wrote that K application tends to reduce fungal diseases. In addition, the high content of humic acids (10.3% of dry matter) could explain the positive effect of the obtained CT on plant growth promotion and resistance. Besides, the application of humic acids enhances phenolic and flavonoid accumulation and improves yields in *Cichorium intybus* plants [40]. Several authors have linked humic substances as compounds having an auxin-like activity that induces nitrate metabolism for plant growth [41,42]. In addition, Morales-Corts et al. [9] explained that the growth effect of CT could be not only explained by humic acid content, but also by nutrient concentration, phytohormones, and the presence of beneficial microorganisms. Therefore, the microbiota in a CT also play a fundamental role in the ability of these extracts to suppress plant diseases or promote growth [13]. Bacterial communities (total aerobic bacteria and N-fixing bacteria) were the predominant microorganisms in this aerated CT. These results follow Kim et al. [43], who observed similar population densities of culturable bacteria in the mixture of oriental medicinal herb compost and vermicompost tea during incubation. Furthermore, Li et al. [44] showed analogous results in microbial populations in maize straw CT, having the most total bacteria microbial populations, followed by actinomycetes and total fungi. It is clear that microbial communities found in a CT can be indirectly responsible for growth improvement with respect to the control because of their implication for the availability of nutrients as previously suggested by Hamid et al. [19], especially due to K^+ content in the studied CT.

Concerning the inhibitory effect of the CT on the studied fungal pathogens, in vitro assay results showed that the application of CT to PDA (1:10) significantly decreased the growth rate of *R. solani* and *P. capsici* in comparison with the control treatment. According to these results, Dionne et al. [45] showed that the CT significantly reduced the growth of the mycelium of *R. solani*. This reduction seems to be related to the high microbiota population, since in our experiment the sterilized CT did not show this direct suppressive effect (data not shown). Hence, CT-enriched media displayed a large number of microorganisms that could directly repress fungi growth. In addition to this statement, several authors have previously written that microbial populations are required to determine the suppressive effect of the CT [36,46]. Moreover, the CT contains different species of the genera *Trichoderma, Penicillium, Aspergillus, Bacillus, Enterobacter, Rhizobacteria* or *Pseudomonas* spp., among others, which could have the ability to control pathogens and stimulate plant growth [47–50].

The effect of a CT on plant growth and resistance against *P. capsici* and *R. solani* was tested. The plants grown under CT treatment (T1) displayed the best results in biomass and in the total and average weight of fruit. These results are in concordance with those indicated by Ros et al. [51], who pointed that the foliar application of CT increases the yield and quality of baby spinach plants. This could be explained by the contribution of nutrient content, humic acids, and microbial population in growth promotion [52]. In addition to this statement, Reeve et al. [53] indicated that the nutrient content in a CT could supplement or substitute for other types of fertilizers.

Moreover, it should be pointed out that the CT application carried out in this trial does not provide all the necessary nutrients for the pepper crop, since the control plants showed nutritional deficiencies from the middle of the cycle. Therefore, a supplementary fertilization is required to cover the nutritional requirements of pepper plants to get a high yield. Therefore, the combination of mineral and organic components should be considered to reduce the chemical resources application. In relation to this statement, several authors tested the combination compost or mineral solubilizing microorganism with fertilizers as a good solution to increase the availability of nutrients [54–56].

It is relevant to the point that the first flower of the CT-treated plants (T1, T2, T3, T6, and T7) started opening one week before those without CT treatment (Table 3), which indicates that application of the CT advanced the crop cycle. Similar results were observed in potato plants treated with garden waste CT [57]. Furthermore, pepper plants inoculated with *P. capsici* one week after transplanting (T5) showed a reduction in fruit weight as well as in stem and root dry weight compared to the control treatment. These plants also displayed a higher grade of pathogen incidence (black roots, narrow neck, thin stem, leaf decay and very pronounced chlorosis). In addition, plants treated with CT and inoculated with *P. capsici* (T3 and T7) showed a reduction in pathogen incidence. They increased growth parameters (total fruit weight, average fruit weight, dry weight of the root, and aerial part and total biomass). This fact may be due to the microbial populations of CT since

Ezziyyani et al. [58] previously revealed that both *Trichoderma harzianum* and actinobacteria effectively reduce the attack of *Phytophthora capsici*.

Moreover, regarding the disease produced by *R. solani*, plants inoculated with this pathogen one week after transplanting (T4) displayed grade 2 pathogen incidence and a reduction in root and stem weight relative to the control. Interestingly, as for *P. capsici*, plants presented less pathogen damage when inoculated with *R. solani* four weeks after transplanting (T8) with respect to those inoculated one week after transplanting (T4). This point led us to conclude that CT application had a suppressive effect on the attack of *R. solani* as shown in in vitro cultures. Moreover, several authors previously studied the effect of different CTs on *R. solani* control in different plant species [59,60]. In this line, Morales-Corts et al. [9] confirmed the effect of green waste CT on the control of this pathogen in potato plants. This fact is directly related to the presence of *Trichoderma* spp., among other microbial antagonists and parasitic or antibiotic-producer microorganisms [46,61], and to indirect mechanisms inducing systemic resistance in plants [19]. In this sense, the addition of PGPR may enhance auxin production in the rhizosphere which is linked to developing stress tolerance in plants. Consequently, the addition of CT could minimize stress in the pepper plants caused by pathogens, especially when the application is carried out before an attack.

Altogether, indications are that this CT could be considered as a green alternative to supplement, or substitute for, other types of fertilizers and pesticides. Since pepper is one of the most important human food crops, management strategies using CT obtained from green waste (following the Sustainable Development Goals of the United Nations to achieve a better and more sustainable future [62]) could be relevant in integrated and sustainable agriculture.

5. Conclusions

This study confirmed the positive effect of CT application (based on green waste) on total biomass and pepper fruit production and for its direct implication for the control of pathogen development by causing a reduction of both *P. capsici* and *R. solani* in in vitro and in vivo essays. The studied extract presented relevant content in NO_3^-, K_2O, humic acids, and microorganisms, which seem to play a synergic role in plant growth and protection. Microbiota in the CT induced a reduction in the relative growth rate of both *P. capsici* and *R. solani* (31.7% and 38.0%, respectively) in in vitro assays compared to the control. Moreover, the effect on pathogen control was significant in *P. capsici* early attacks in the in vivo assays. The CT application produced a 36% increase in biomass with respect to the control, and plants treated with the CT started flowering one week before those that were not treated, thus indicating that the application of the CT advanced the crop cycle. Despite its bio-stimulant action on plant growth, CT application in this trial did not provide all the necessary nutrients for the pepper crop. Therefore, supplementary fertilization is required to cover the nutritional requirements to obtain a high yield. Further studies should be conducted under field conditions and combined with other fertilizers.

Author Contributions: Conceptualization, M.R.M.-C. and M.B.S.-F.; Methodology, M.R.M.-C., R.P.-S. and M.B.S.-F.; Software, A.I.G.-H., M.Á.G.-S. and M.R.M.-C.; Validation, M.R.M.-C. and A.I.G.-H.; Investigation, M.R.M.-C., M.Á.G.-S., A.I.G.-H., R.P.-S. and M.B.S.-F.; Data Curation, M.R.M.-C., M.Á.G.-S., A.I.G.-H., R.P.-S. and M.B.S.-F.; Writing—Original Draft Preparation, A.I.G.-H. and M.R.M.-C.; Writing—Review & Editing, A.I.G.-H., M.R.M.-C. and R.P.-S.; Project Administration, M.R.M.-C.; Funding Acquisition, M.R.M.-C. All authors have read and agreed to the published version of the manuscript.

Funding: This research was supported by Fertiberia and Salamanca University 83 Contract Agreement (Grow-in inductive fertilization).

Institutional Review Board Statement: Not applicable.

Acknowledgments: The authors thank Neiker (Basque Center of Agricultural Research) for the *Phytophthora capsici* strain pathogen and to the Regional Center for pest and Diseases Diagnosis, Junta de Castilla y León, Spain for the *Rhizoctonia solani* strain pathogen.

Conflicts of Interest: The authors declare no conflict of interest.

References

1. Zaller, J.G. Vermicompost as a substitute for peat in potting media: Effects on germination, biomass allocation, yields and fruit quality of three tomato varieties. *Sci. Hortic.* **2007**, *112*, 191–199. [CrossRef]
2. Bernal, M.P.; Alburquerque, J.A.; Moral, R. Composting of animal manures and chemical criteria for compost maturity assessment. A review. *Bioresour. Technol.* **2009**, *100*, 5444–5453. [CrossRef]
3. Benito, M.; Masaguer, A.; De Antonio, R.; Moliner, A. Use of pruning waste compost as a component in soilless growing media. *Bioresour. Technol.* **2005**, *96*, 597–603. [CrossRef]
4. Darzi, M.T.; Seyedhadi, M.H.; Rejali, F. Effects of the application of vermicompost and phosphate solubilizing bacterium on the morphological traits and seed yield of anise (*Pimpinella anisum* L.). *J. Med. Plants Res.* **2012**, *6*, 215–219.
5. Moretti, S.M.L.; Bertoncini, I.B.; Abreu-Junior, C.H. Composting sewage sludge with green waste from tree pruning. *Sci. Agric.* **2015**, *72*, 432–439. [CrossRef]
6. Marin, F.; Santos, M.; Dianez, F.; Carretero, F.; Gea, F.J.; Yau, J.A.; Navarro, M.J. Characters of compost teas from different sources and their suppressive effect on fungal phytopathogens. *World J. Microbiol. Biotechnol.* **2013**, *29*, 1371–1381. [CrossRef] [PubMed]
7. Markakis, E.A.; Fountoulakis, M.S.; Daskalakis, G.C.; Kokkinis, M.; Ligoxigakis, E.K. The suppressive effect of compost amendments on *Fusarium oxysporum* f.sp. *radicis-cucumerinum* in cucumber and *Verticillium dahliae* in eggplant. *Crop Prot.* **2016**, *79*, 70–79. [CrossRef]
8. Al-Dahmani, J.H.; Abbasi, P.A.; Miller, S.A.; Hoitink, H.A.J. Suppression of bacterial spot of tomato with foliar sprays of compost extracts under greenhouse and field conditions. *Plant Dis.* **2003**, *87*, 913–919. [CrossRef] [PubMed]
9. Morales-Corts, M.R.; Pérez-Sánchez, R.; Gómez-Sánchez, M.A. Efficiency of garden waste compost teas on tomato growth and its suppressiveness against soilborne pathogens. *Sci. Agric.* **2018**, *75*, 400–409. [CrossRef]
10. Ingham, E.R. What is compost tea? Part 1. *BioCycle* **1999**, *40*, 74–75.
11. Martin, C.C.G.; Dorinvil, W.; Brathwaite, R.A.I.; Ramsubhag, A. Effects and relationships of compost type, aeration and brewing time on compost tea properties, efficacy against *Pythium ultimum*, phytotoxicity and potential as a nutrient amendment for seedling production. *Biol. Agric. Hortic.* **2012**, *28*, 185–205. [CrossRef]
12. Mengesha, W.K.; Powel, S.M.; Evans, K.J.; Barry, K.M. Diverse microbial communities in non-aerated compost teas suppress bacterial wilt. *World J. Microbiol. Biotechnol.* **2017**, *33*, 49. [CrossRef]
13. De Corato, U. Agricultural waste recycling in horticultural intensive farming systems by on-farm composting and compost-based tea application improves soil quality and plant health: A review under the perspective of a circular economy. *Sci Total Environ.* **2020**, *738*, 139840. [CrossRef] [PubMed]
14. Zaccardelli, M.; Sorrentino, R.; Caputo, M.; Scotti, R.; De Falco, E.; Pane, C. Stepwise-Selected *Bacillus amyloliquefaciens* and *B. subtilis* Strains from Composted Aromatic Plant Waste Able to Control Soil-Borne Diseases. *Agriculture* **2020**, *10*, 30. [CrossRef]
15. Castano, R.; Borrero, C.; Aviles, M. Organic matter fractions by SP-MAS 13C NMR and microbial communities involved in the suppression of Fusarium wilt in organic growth media. *Biol. Control* **2011**, *58*, 286–293. [CrossRef]
16. Siddiqui, Y.; Sariah, M.; Ismail, M.R.; Asgar, A. *Trichoderma*-fortified compost extracts for the control of Choanephora wet rot in okra production. *Crop Prot.* **2008**, *27*, 385–390. [CrossRef]
17. Pant, A.P.; Radovich, T.J.; Hue, N.V.; Paull, R.E. Biochemical properties of compost tea associated with compost quality and effects on pak choi growth. *Sci. Hortic.* **2012**, *148*, 138–146. [CrossRef]
18. Ilangumaran, G.; Smith, D.L. Plant growth promoting rhizobacteria in amelioration of salinity stress: A systems biology perspective. *Front. Plant Sci.* **2017**, *8*, 1768. [CrossRef]
19. Hamid, S.; Ahmad, I.; Akhtar, M.J.; Iqbal, M.N.; Shakir, M.; Tahir, M.; Rasool, A.; Sattar, A.; Khalid, M.; Ditta, A.; et al. *Bacillus subtilis* Y16 and biogas slurry enhanced potassium to sodium ratio and physiology of sunflower (*Helianthus annuus* L.) to mitigate salt stress. *Environ. Sci. Pollut Res. Int.* **2021**. Online ahead of print. [CrossRef]
20. Martin, C.C.S. Potential of compost tea for suppressing plant diseases. *CAB Rev.* **2014**, *9*, 1–38.
21. Pane, C.; Celano, G.; Villecco, D.; Zaccardelli, M. Control of *Botrytis cinerea*, *Alternaria alternata* and *Pyrenochaeta lycopersici* on tomato with whey compost-tea applications. *Crop Prot.* **2012**, *38*, 80–86. [CrossRef]
22. Shaban, H.; Fazeli-Nasab, B. An Overview of the Benefits of Compost tea on Plant and Soil Structure. *Adv. Biores.* **2015**, *6*, 154–158.
23. Food and Agriculture Organization (FAO). Faostat: Agriculture Data. 2019. Available online: http://www.fao.org/faostat/en/#data/QC/visualize (accessed on 5 December 2020).
24. Sundaramoorthy, S.; Raguchander, T.; Ragupathi, N.; Samiyappan, R. Combinatorial effect of endophytic and plant growth promoting rhizobacteria against wilt disease of *Capsicum annuum* L. caused by *Fusarium Solani*. *Biol. Control* **2012**, *60*, 59–67.
25. Kim, E.H.; Lee, S.Y.; Baek, D.Y.; Park, S.Y.; Lee, S.G.; Ryu, T.H.; Lee, S.K.; Kang, H.J.; Kwon, O.H.; Kil, M.; et al. A comparison of the nutrient composition and statistical profile in red pepper fruits (*Capsicums annuum* L.) based on genetic and environmental factors. *Appl. Biol. Chem.* **2019**, *62*, 48. [CrossRef]

26. Barchenger, D.W.; Lamour, K.H.; Bosland, P.W. Challenges and Strategies for Breeding Resistance in *Capsicum annuum* to the Multifarious Pathogen, *Phytophthora Capsici*. *Front Plant Sci.* **2018**, *9*, 628. [CrossRef]
27. Sherf, A.F.; MacNab, A.A. *Vegetable Diseases and Their Control*, 2nd ed.; Wiley: New York, NY, USA, 1986; pp. 334–337.
28. Mannai, S.; Jabnoun-Khiareddine, H.; Nasraoui, B.; Daami-Remadi, M. Rhizoctonia Root Rot of Pepper (*Capsicum annuum*): Comparative Pathogenicity of Causal Agent and Biocontrol Attempt using Fungal and Bacterial Agents. *J. Plant Pathol. Microbiol.* **2018**, *9*, 1000431.
29. Morales-Corts, M.R.; Gómez-Sánchez, M.A.; Pérez-Sánchez, R. Evaluation of green/pruning wastes compost and vermicompost, slumgum compost and their mixes as growing media for horticultural production. *Sci. Hortic.* **2014**, *172*, 155–160. [CrossRef]
30. Wickerham, L.J. Taxonomy of yeasts. *U.S. Dept. Agric. Tech. Bull.* **1951**, *1029*, 1–56.
31. Sanchis Solera, J. Comparación entre los diversos medios de cultivo comerciales para aislamiento de hongos (levaduras y mohos). In *Proyecto Microkit 1999 Para Optimizar la Sensibilidad de los Parámetros de Muestras Microbiológicos del aire*; Laboratorios Microkit, S.L.: Madrid, Spain, 1996.
32. Stella, M.; Suhaimi, M. Selection of suitable growth medium for free-living diazotrophs isolated from compost. *J. Trop. Agric. Fd. Sc.* **2010**, *38*, 211–219.
33. Vargas-Gil, S.; Pastor, S.; March, G.J. Quantitative isolation of biocontrol agents *Trichoderma* spp., *Gliocladium* spp. and actinomycetes from soil with culture media. *Microbiol. Res.* **2006**, *164*, 196–205. [CrossRef]
34. Köller, W.; Wilcox, W.F. Evidence for the predisposition of fungicide-resistant isolates of venturia inaequalis to a preferential selection for resistance to other fungicides. *Phytopathology* **2001**, *91*, 776–781. [CrossRef] [PubMed]
35. Holmes, K.A.; Benson, D.M. Evaluation of *Phytophthora parasitica* var. nicotianae for biocontrol of *Phytophthora parasitica* on *Cathar. Roseus*. *Plant Dis.* **1994**, *78*, 193–199. [CrossRef]
36. Scheuerell, S.J.; Mahaffee, W.F. Compost tea as a container medium drench for suppressing seedling damping-off caused by *Pythium ultimum*. *Phytopathology* **2004**, *94*, 1156–1163. [CrossRef]
37. Siddiqui, Y.; Meon, S.; Ismai, R.; Rahmani, M. Bio-potential of compost tea from agro-waste to suppress *Choanephora cucurbitarum* L. the causal pathogen of wet rot of okra. *Biol. Control* **2009**, *49*, 38–44. [CrossRef]
38. Naidu, Y.; Meon, S.; Kadir, J.; Siddiqui, Y. Microbial starter for the enhancement of biological activity of compost tea. *Int. J. Agric Biol.* **2010**, *12*, 51–56.
39. Amtmann, A.; Troufflard, S.; Armengaud, P. The effect of potassium nutrition on pest and disease resistance in plants. *Physiol. Plant.* **2008**, *133*, 682–691. [CrossRef]
40. Gholami, H.; Saharkhiz, M.J.; Fard, F.R.; Ghani, A.; Nadaf, F. Humic acid and vermicompost increased bioactive components, antioxidant activity and herb yield of Chicory (*Cichorium intybus* L.). *Biocatal. Agric. Biotechnol.* **2018**, *14*, 286–292. [CrossRef]
41. Eyheraguibel, B.; Silvestre, J.; Morard, P. Effects of humic substances derived from organic waste enhancement on the growth and mineral nutrition of maize. *Bioresour Technol.* **2008**, *99*, 4206–4212. [CrossRef]
42. Muscolo, A.; Sidari, M. Carboxyl and phenolic humic fractions affect callus growth and metabolism. *Soil Sci. Soc. Am. J.* **2009**, *73*, 1119–1129. [CrossRef]
43. Kim, M.J.; Shim, C.K.; Kim, Y.K.; Hong, S.J.; Park, J.H.; Han, E.J.; Kim, J.H.; Kim, S.C. Effect of Aerated Compost Tea on the Growth Promotion of Lettuce, Soybean, and Sweet Corn in Organic Cultivation. *Plant Pathol. J.* **2015**, *31*, 259–268. [CrossRef]
44. Li, X.; Wang, X.; Shi, X.; Wang, Q.; Li, X.; Zhang, S. Compost tea-mediated induction of resistance in biocontrol of strawberry Verticillium wilt. *J. Plant Dis. Prot.* **2020**, *127*, 257–268. [CrossRef]
45. Dionne, A.; Tweddell, R.J.; Antoun, H.; Avis, T.J. Effect of non-aerated compost teas on damping-off pathogens of tomato. *Can. J. Plant Pathol.* **2012**, *34*, 51–57. [CrossRef]
46. Diánez, F.; Santos, M.; Tello, J.C. Suppressive effects of grape marc compost on phytopathogenic oomycetes. *Arch. Phytopathol. Plant Prot.* **2007**, *40*, 1–18. [CrossRef]
47. Phae, C.G.; Shoda, M. Expression of the suppresive effects of Bacillus sibtilis on phytopathogens in inoculated compost. *J. Ferment Bioeng* **1990**, *70*, 409–414. [CrossRef]
48. Scheuerell, S.J.; Mahaffee, W.F. Compost Tea: Principles and Prospects for Plant Disease Control. *Compost. Sci. Util.* **2002**, *10*, 313–338. [CrossRef]
49. Sylvia, E.W. The effect of compost extract on the yield of strawberries and severity of *Botrytis Cinerea*. *J. Sustain Agr.* **2004**, *25*, 57–68.
50. Ingham, E.R. *The Compost Tea Brewing Manual*, 5th ed.; Soil Food International Inc.: Corvallis, OR, USA, 2005.
51. Ros, M.; Hurtado-Navarro, M.; Giménez, A.; Fernández, J.A.; Egea-Gilabert, C.; Lozano-Pastor, P.; Pascual, J.A. Spraying Agro-Industrial Compost Tea on Baby Spinach Crops: Evaluation of Yield, Plant Quality and Soil Health in Field Experiments. *Agronomy* **2020**, *10*, 440. [CrossRef]
52. Marín, F.; Diánez, F.; Santos, M.; Carretero, F.; Gea, F.J.; Castañeda, C.; Navarro, M.J.; Yau, J.A. Control of *Phytophthora capsici* and *Phytophthora parasitica* on pepper (*Capsicum annuum* L.) with compost teas form different sources, and their effects on plant growth promotion. *Phytopathol. Mediterr.* **2014**, *53*, 216–228.
53. Reeve, J.R.; Carpenter-Boggs, L.; Reganold, J.P.; York, A.L.; Brinton, W.F. Influence of biodynamic preparations on compost development and resultant compost extracts on wheat seedling growth. *Bioresour. Technol.* **2010**, *101*, 5658–5666. [CrossRef]

54. Ditta, A.; Imtiaz, M.; Mehmood, S.; Rizwan, M.S.; Mubeen, F.; Aziz, O.; Qian, Z.; Ijaz, R.; Tu, S. Rock phosphate enriched organic fertilizer with phosphate solubilizing microorganisms improves nodulation, growth and yield of legumes. *Comm. Soil Sci. Plant Anal* **2018**, *49*, 2715–2725. [CrossRef]

55. Ditta, A.; Muhammad, J.; Imtiaz, M.; Mehmood, S.; Qian, Z.; Tu, S. Application of rock phosphate enriched composts increases nodulation, growth and yield of chickpea. *Int. J. Recycl. Org. Waste Agric.* **2018**, *7*, 33–40. [CrossRef]

56. Hussain, A.; Zahir, Z.A.; Ditta, A.; Tahir, M.U.; Ahmad, M.; Mumtaz, M.Z.; Hayat, K.; Hussain, S. Production and Implication of Bio-Activated Organic Fertilizer Enriched with Zinc-Solubilizing Bacteria to Boost up Maize (*Zea mays* L.) Production and Biofortification under Two Cropping Seasons. *Agronomy* **2020**, *10*, 39. [CrossRef]

57. López-Martín, J.J.; Morales-Corts, M.R.; Pérez-Sánchez, R.; Gómez-Sánchez, M.A. Efficiency of garden waste compost teas on potato growth and its suppressiveness against *Rhizoctonia*. *Agric. For.* **2018**, *64*, 7–14.

58. Ezziyyani, M.; Requena, M.E.; Pérez-Sánchez, C.; Candela, M.E. Efecto del sustrato y la temperatura en el control biológico de *Phytophthora capsici* en pimiento (*Capsicum annuum* L.). *An. de Biol.* **2005**, *27*, 119–126.

59. Tateda, M.; Yoneda, D.; Sato, Y. Effects of Compost Tea Making from Differently Treated Compost on Plant Disease Control. *J. Wetl. Res.* **2007**, *9*, 91–98.

60. Pane, C.; Piccolo, A.; Spaccini, R.; Celano, G.; Villecco, D.; Zaccardelli, M. Agricultural waste-based composts exhibiting suppressivity to diseases caused by the phytopathogenic soil-borne fungi *Rhizoctonia solani* and *Sclerotinia minor*. *Appl. Soil Ecol.* **2013**, *65*, 43–51. [CrossRef]

61. Krause, M.S.; Madden, L.V.; Hoitink, H.A.J. Effect of potting mix microbial carrying capacity on biological control of Rhizoctonia damping-off of radish and Rhizoctonia crown and root rot of poinsettia. *Phytopathology* **2001**, *91*, 1116–1123. [CrossRef] [PubMed]

62. United Nations (UN). The 17 Goals. Department of Economic and Social Affairs. 2021. Available online: https://sdgs.un.org/goals (accessed on 8 February 2021).

MDPI

St. Alban-Anlage 66

4052 Basel

Switzerland

Tel. +41 61 683 77 34

Fax +41 61 302 89 18

www.mdpi.com

Agronomy Editorial Office

E-mail: agronomy@mdpi.com

www.mdpi.com/journal/agronomy

www.ingramcontent.com/pod-product-compliance
Lightning Source LLC
LaVergne TN
LVHW070613170726
843515LV00004B/68